CHEMICAL AND METALLURGICAL THERMODYNAMICS

CHEMICAL AND METALLURGICAL THERMODYNAMICS

Krishna Kant Prasad

Hem Shanker Ray

K.P. Abraham

PUBLISHING FOR ONE WORLD

NEW AGE INTERNATIONAL (P) LIMITED, PUBLISHERS

New Delhi • Bangalore • Chennai •Cochin • Guwahati • Hyderabad

Jalandhar • Kolkata • Lucknow • Mumbai • Ranchi

Visit us at **www.newagepublishers.com**

Published by New Age International (P) Ltd., Publishers
First Edition: 2007
Reprint: 2012

BRANCHES

- **Bangalore** 37/10, 8th Cross (Near Hanuman Temple), Azad Nagar, Chamarajpet, Bangalore-560 018
  Tel.: (080) 26756823, Telefax: 26756820, E-mail: bangalore@newagepublishers.com
- **Chennai** 26, Damodaran Street, T. Nagar, Chennai-600 017
  Tel.: (044) 24353401, Telefax: 24351463, E-mail: chennai@newagepublishers.com
- **Cochin** CC-39/1016, Carrier Station Road, Ernakulam South, Cochin-682 016
  Tel.: (0484) 2377004, Telefax: 4051303, E-mail: cochin@newagepublishers.com
- **Guwahati** Hemsen Complex, Mohd. Shah Road, Paltan Bazar, Near Starline Hotel, Guwahati-781 008
  Tel.: (0361) 2513881, Telefax: 2543669, E-mail: guwahati@newagepublishers.com
- **Hyderabad** 105, 1st Floor, Madhiray Kaveri Tower, 3-2-19, Azam Jahi Road, Nimboliadda, Hyderabad-500 027
  Tel.: (040) 24652456, Telefax: 24652457, E-mail: hyderabad@newagepublishers.com
- **Kolkata** RDB Chambers (Formerly Lotus Cinema) 106A, 1st Floor, S.N. Banerjee Road, Kolkata-700 014
  Tel.: (033) 22273773, Telefax: 22275247, E-mail: kolkata@newagepublishers.com
- **Lucknow** 16-A, Jopling Road, Lucknow-226 001
  Tel.: (0522) 2209578, 4045297, Telefax: 2204098, E-mail: lucknow@newagepublishers.com
- **Mumbai** 142C, Victor House, Ground Floor, N.M. Joshi Marg, Lower Parel, Mumbai-400 013
  Tel.: (022) 24927869, Telefax: 24915415, E-mail: mumbai@newagepublishers.com
- **New Delhi** 22, Golden House, Daryaganj, New Delhi-110 002
  Tel.: (011) 23262368, 23262370, Telefax: 43551305, E-mail: sales@newagepublishers.com

ISBN (10) : 81-224-1978-X
ISBN (13) : 978-81-224-1978-8

₹ 295.00

C-12-06-6402

Printed in India at Glorious Printers, Delhi.
Typeset at Akriti Graphics, Delhi.

PUBLISHING FOR ONE WORLD
NEW AGE INTERNATIONAL (P) LIMITED, PUBLISHERS
7/30 A, Daryaganj, New Delhi-110002
Visit us at **www.newagepublishers.com**

Preface

My first interaction with the science of thermodynamics was not a pleasant one. And I have reasons to presume that for most science students the experience is similar. Soon after, I started patting myself on my back since I felt that I could understand the subject better than my classmates, and with a little help from my teachers I should be able to master the subject. But I found my teachers also woefully lacking in clarity of concepts. Thereafter, my enthusiasm for Physical Chemistry and Thermodynamics remained subsided until I became a student of metallurgy. The practical applicability of thermodynamics in metallurgical processes, and the clarity of concepts taught to us by Prof. K.P. Abraham, rekindled my interest so much so that, at that time I decided to make it my career. My Ph.D work was related to thermodynamic properties and I undertook teaching Metallurgical Thermodynamics, first at the Banaras Hindu University, and thereafter at Jamshedpur Technical Institute.

After joining steel industry, I became somewhat lethargic in my academic pursuits. Yet it was always in the back of my mind that some means should be devised so that a student of science should not find his/her first interaction with Thermodynamics an unpleasant one. I had conceived of a book, which could present Thermodynamics in a concise but simple way. The present book is an attempt in that direction.

But my experience has been that in trying to make it concise we do not necessarily make it simple. In the concise or precise form, it looks too mathematical to be easily understood. Thus, this attempt has been somewhat of a compromise.

This compromise has resulted in the addition of two introductory chapters covering the historical perspective and the aspect of feasibility. The book in exact sense starts from chapter 3. Since for understanding the introductory chapters 1 and 2, some prior knowledge of Chemical and Metallurgical Thermodynamics is needed, a beginner should skip these two chapters during the first reading.

Chapters 3 to 7 comprise the basic treatment of "Classical Thermodynamics"- a term explained at the end of chapter 3. Chapters 8 to 10 are, in a way, appendices to classical thermodynamics. But, since lines of demarcation with other forms of thermodynamics have become thin, we have not named them as appendices.

The beauty of thermodynamics is (here I am not alone) its universality at macro level. They say that all laws may change but laws of Thermodynamics would never change. The other important aspect of Thermodynamics is that it is based on very common observations and thus it is based on very simple truths. But after all the Theory of Relativity is also based on two very simple facts!!!

KKP

Symbols, Abbreviations and Notes

α = Degree of Reduction

$\gamma = C_p/C_v$ for gases, *also see below*

γ = Activity coefficient (in the context of solution thermodynamics)

η = Efficiency

∂ = Partial differential

δ, Δ = Change in property

ε = Energy level of an ensemble (in the context of statistical thermodynamics)

$\in$ = Interaction coefficient

θ = Temperature

A= Work function, Helmholtz Free Energy

°C = Degrees Centigrade (temperature)

C_p = Heat Capacity at constant pressure

C_v = Heat Capacity at constant volume

d = Differential

e = Henrian interaction coefficient, *also see below*

e = Natural exponential number/Euler's number, approx. value 2.71828

EMF = Electromotive Force

E = Internal Energy, *also see below*

E = EMF, or Electromotive Force, or Cell Potential (In the context of galvanic cells); *also see below*

E= Activation Energy (In the context of reaction kinetics)

f= Fugacity, *also see below*

f= Henrian activity coefficient

F = Free Energy, Gibbs Free Energy

F_a = Farady Constant

g = Gas

H = Enthalpy, Heat Content

k = Boltzmann's Constant (In the context of Statistical Thermodynamics)

°K = Kelvin (Absolute temperature in Centigrade scale)

l = Liquid

n = Number of Moles (gram-moles, kilogram-moles, pound-moles, etc.); *also see below*

n = Number of molecules, atoms, sites or micro-states

n_A, n_B, etc. = Number of moles of constituent 'A', 'B', etc.

N = Mole (or Atom) Fraction

N_a = Avogadro's Number

N_A, N_B, etc. = Mole Faction of constituent 'A', 'B', etc.

P = Pressure

Q, q = Heat (Transferred – either into or out of the system)

R = Gas Constant

°R = Degrees Rankine (Absolute temperature in Fahrenheit scale)

s = Solid

S = Entropy

Sec. = Section

T = Temperature (Absolute)

U = Energy (in general sense)

V = Volume

W = Work Done; *also see below*

W = Number of Micro-states (In the context of Statistical Thermodynamics)

Notes:

- For arithmetic operation *(Star) has been used to denote multiplication to avoid confusion with alphabet x. Also / (slash) has been used to denote division to avoid increasing line space.
- Italics in text generally denote the reference in the Bibliography section.
- CGS units have been used in the text and not SI units, as is the current trend. One important conversion factor, which is required for linking SI and CGS systems is that, one calorie equals 4.186 joules.

List of Figures

List of Tables

Contents

Chapter- 1

Historical Perspective of Thermodynamics

A study of historical perspective of thermodynamics is helpful in understanding the underlying concepts better. But understanding history requires some knowledge regarding the laws of thermodynamics. A beginner is therefore advised to skip this section during first reading.

Inter-convertibility of energy and the concept of temperature are used in many other aspects of science. Hence, postulations of the Zeroeth and the First laws of Thermodynamics do not signal the start of the science of thermodynamics. We can say that the birth of thermodynamics was signalled in the year 1824 when Carnot *(Magie 1899)*, who was investigating steam engine, with a view to increase its efficiency, stated his principle (theorem) from which the second law of thermodynamics evolved. He said that the 'motive (work) power of heat' depends solely on the 'temperatures of the bodies between which the transfer of heat occurs'. Thomson *(1882)* spent many years trying to reconcile this statement with the 'Mechanical Equivalent of Heat' postulated by Joule *(1884)* and, probably assisted by the publication of Clausius *(Magie 1899)*, concluded that the two phenomena are not contradictory but are two different laws. After the clarifications of Clausius and Thomson, clarity could be found in the science of thermodynamics from about 1850 onwards.

Clausius himself, later on, introduced the concept of entropy. Studies on entropy led to the evolution of the 'Third Law of Thermodynamics' in 1906 by Nernst *(1907)*. This was many times debated and revised till 1912, when Plank *(1927)* went back to practically the same statement as that of Nernst.

With this, the development of Classical Thermodynamics was complete. Or was it? In as late as the middle of twentieth century it was argued and agreed that the 'principle of unattainability of the absolute zero' is synonymous with the third law statement of entropy *i.e.*, 'entropy of a crystalline substance is zero at absolute zero temperature'.

We have used the term 'Classical Thermodynamics' – a science which has evolved out of the study of heat and its capacity to do work. This was the thermodynamics applied to heat engines – a physical system. But the concepts were found to apply to chemical systems (Chemical Thermodynamics) as well as high temperature chemical reactions involving metals, its oxides and other compounds (Metallurgical Thermodynamics). The evolution of concepts of 'Entropy' and 'Free Energy', which determined 'Randomness' and 'Feasibility' respectively, led to its application in fine-tuning the already existing processes (e.g., synthesis of ammonia from nitrogen and hydrogen, sponge iron making in rotary kiln, etc.) and in the development of newer processes.

The concept of entropy and its dependence on randomness, led to the interpretation of thermodynamic properties in terms of atomic or molecular arrangements. Earlier also, there were attempts to correlate, rather logically, the temperature with molecular motion. But till the evolution of entropy concept, thermodynamics studied properties of system at macro-level only. The latter interpretations with atomic and molecular arrangements came to be discussed under 'Statistical Thermodynamics.'

The thermodynamic driving force of a reaction – the free energy change and the equilibrium constant – was latter used extensively in study of kinetics of reaction, since most of the process developers wanted to know under what conditions a reaction can be driven faster – at as low a temperature as possible. Higher temperatures not only demanded extra energy, but caused problems which could be reduced or even eliminated, at lower temperatures. Estimation of such driving force in complex systems required very large number of calculations, something that was made possible with the advent of high-capacity computers. A new branch of 'Computational Thermodynamics' is now evolving.

Chapter- 2

Thermodynamics and Feasibility

2.1. WHAT MAKES A REACTION FEASIBLE?

Scientists, starting with alchemists, have since long been pondering over the reasons for what makes a particular chemical reaction feasible or possible. We have used the word *feasible* here, which means 'capable of being done, accomplished or carried out.' We know through commonsense that a mixture of hydrogen and oxygen gases at room temperature or below would not react but a reaction between them is very much feasible. Only the rate of reaction is too small to be measurable. Please do not try to prepare such a mixture. It could have a disastrous result (see Sec. 4.5.7).

Scientists sensed that since most of the reactions occurred with liberation of heat, and intensity of reaction often matched the intensity of heat liberated, the extent of heat liberation had some relationship with the driving force behind the reaction. Further, it was noticed that, for high intensities of reaction, it was more difficult to revert to the reactants. The obvious conclusion, that the extent of heat liberation was a measure of the feasibility of reaction, could easily be disproved, but there was a definite linkage between the two as subsequent developments proved.

2.2. HEAT AND WORK – ARE THEY SYNONYMOUS? WHAT DOES THERMODYNAMICS SAY?

Parallely, the science of thermodynamics developed, which was based on the observations on the properties of heat as distinct from other forms of energy. The science of thermodynamics deals with transformation of other forms of energy to heat, and from heat to other forms. The term 'dynamics' in thermodynamics refers to 'transformation' rather than transfer. The science of heat transfer is covered under the science of Thermophysics.

It was easily seen that friction generates heat and it was a case of mechanical work being converted into heat. Heat was indestructible but could 'easily' be transformed to and from other forms of energy (First law of Thermodynamics – Sec. 4.4). It could also be seen that a lighted candle could boil water in a test tube in matter of minutes but could not do so even in hours for a bucketful of water even though a lot more (quantity of) heat was supplied to the bucket.

Obviously, there is a level of heat (temperature) much like the level of water. Thus, just as in interconnected containers water would settle at the same level, similarly if two bodies are in thermal equilibrium with a third body they would be in thermal equilibrium with each other (Zeroeth Law of thermodynamics – Sec. 4.1.1). The importance of this Zeroeth Law was realised well after the development of science of thermodynamics even though the concept was used throughout without formally recognising it. This Zeroeth Law of thermodynamics emphasises the fluid nature of heat – a property also exhibited by most of the other forms of energy (*e.g.*, Electrical Potential in a conductor).

2.3. HEAT ENGINES

Scientists also noticed that when they tried to convert heat into work, the matter was not as simple as converting work into heat. A machine, which would take heat and convert into work (a heat engine), would necessarily have to discard some (lower quality) heat (Second Law of Thermodynamics – Sec. 5.1). In other words, difference between heat and work (and other forms of energy) was emphasised by this law.

2.4. ENTROPY

Even if the heat engine operates in theoretically the most efficient manner, the efficiency of conversion is always considerably less than unity. The extent of conversion is different for different conditions of operation of heat engine. It was noted that with slower rates of operation of the heat engine, the efficiency of conversion of heat into work increased, tending to a maximum when the conversion is carried out at infinitesimally slow speed – something that is not realisable in real processes. Under such a condition, the efficiency of conversion of heat into work depended only on the temperatures of heat source and heat sink (where the lower quality heat is discarded). Thermodynamists termed such an operation as a "Reversible Operation" (see Sec. 4.4.5) where equilibrium conditions are always maintained. This thermodynamically reversible process should not be confused with the reversible changes, either physical (*e.g.*, water changing into ice and back) or chemical (*e.g.*, mercury oxidising into mercuric oxide and back into mercury on heating).

Even though thermodynamically reversible processes cannot be real processes, they have an important significance as they indicate a limit to which real processes may attempt to reach. We may say that "thermodynamically reversible processes" replicate the "real processes" in an "ideal mode".

What happens when such real processes are carried out in an ideal mode? Let us imagine that a reversible heat engine is operating between two temperatures T_1 and T_2, *i.e.*, the heat source is at T_1 and the heat sink is at T_2. In a complete cycle of operation the system of heat engine returns to its original temperature (or original state) after taking, say Q_1 quantity of heat from the source at T_1, discarding Q_2 quantity of heat to the sink at T_2 and carrying out work W on the surroundings, which is algebraically equal to $Q_1 - Q_2$. But if the cycle is not complete, most of the attributes of the closed system of heat engine change, even though equilibrium conditions were maintained throughout.

Is there anything, which remains constant during such an operation *i.e.,* when cycle is not completed and the system has not returned to its original state? Yes, it is the "Entropy" (see Sec. 6.4), which remains constant if the process is carried out in fully-equilibrated condition. For any other condition (*i.e.*, for real processes), the 'Entropy" of system and surroundings taken together always increases.

The property defined by the term "Entropy" was originally coined to explain the difference in properties of heat and other types of energy. Later it was realised that there is a linkage between entropy and the extent of disorder. An increase in the entropy of a system is associated with an increase in disorder.

What was applicable to a heat engine was then successfully adopted for chemical reactions.

2.5. LINKAGE BETWEEN HEAT ENGINES & CHEMICAL REACTIONS

The source of heat in a typical heat engine, such as the Internal or the External Combustion Engine, is the heat evolved by a chemical reaction (burning of a fuel). The chemical heat, when utilised in a heat engine, is partly used in doing work while the balance is discarded to the surroundings. The efficiency of conversion of heat into work may remain limited, but if the engine operates at near equilibrated conditions, the entropy increase can be kept to a minimum. The more the heat engine operates away from the equilibrium conditions the efficiency of conversion decreases and the entropy keeps on increasing. When the heat engine is not operating or is not present, the entire chemical heat would be dissipated in the surroundings and the entropy increase would be even higher.

The concept of Entropy was found to remain valid irrespective of the path the heat took to flow from source to sink, *i.e.*, either through the heat engine or directly from chemical reaction into the atmosphere. Thus, for any real process occurring in real system, there is a net increase in the entropy of system and surroundings taken together.

2.6. ENTROPY AND DISORDER

2.6.1. Inflating a Hydrogen Baloon

It is not very difficult to visualise a thermodynamically reversible process. Suppose we want to inflate a balloon, which we want later to climb up into atmosphere. We connect the balloon to a cylinder containing compressed hydrogen gas. And we allow the gas to pass very slowly from the cylinder to the balloon. The slower this is done, the closer is the inflation process to a thermodynamically reversible process.

Now let us assume that after some time – obviously in infinite time for a reversible process – the balloon is in fully-inflated state and is being ever so slowly being inflated further. How quiet and "orderly" this process is! Suddenly disaster strikes. Some mischief monger pricks the balloon with a pin – or the balloon otherwise gives way. The changes are spontaneous – far removed from the concept of thermodynamic reversibility.

There is a loud noise. Bits and pieces of balloon get scattered all around. The hydrogen gas inside the balloon, which was clearly separated from the air around, gets mixed with air. There probably was a sudden drop in temperature of hydrogen molecules as the pressure on it was released. All this presents a totally chaotic and disorderly picture. No wonder that scientists have associated increased entropy with increased disorder.

2.6.2. Localised Ordering – Dropping Supercooled Ice into Water

As all natural processes are spontaneous processes – all far away from reversible process – entropy of the universe is always increasing. There may be localised ordering process taking place, frequently resulting in local decrease in entropy, but this is always accompanied by a larger increase in entropy elsewhere.

If a piece of super-cooled ice (much below 0 °C) is allowed to fall into water, the layer of water first coming into contact with the ice-piece would freeze and pass into a more orderly state than water. The entropy for this layer has decreased. But this has also caused increase in temperature of the rest of the ice – increased vibration of its molecules – and a larger increase in entropy there.

2.6.3. Boiling of Water, Earthquakes

There could be many more examples of spontaneous process leading to increased disorder. When water boils to give steam, the water molecules pass from a condensed (liquid) phase to a more disorderly gaseous phase. When molten magma below the earth's crust builds up pressure, which the crust cannot sustain, earthquake occurs bringing chaos in the vicinity.

2.6.4. The Ageing Process and Disorder

The human body consists of cells built up of chains and networks of different types of protein and lipid molecules. For younger persons, these chains and networks are very perfect and orderly but as he grows old (although slowly but spontaneously, *i.e.*, irreversibly) the structure develops "defects" or "disorders" signifying an increase in entropy. This particular example needs to be accepted only figuratively as it is difficult to derive a mathematical correlation.

2.6.5. Burning of Fuels

And finally all the energy intensive modern industries are liberating a large quantity of carbon dioxide into atmosphere leading to **global warming**. A huge quantity of carbon, which was in condensed phase earlier, has now joined the much more disorderly gaseous phase and has increased the average carbon dioxide level in the air we breathe from 0.0025% to 0.004% over the last one hundred years. Surely, we are headed for much greater disorder and much greater entropy in future.

2.7. FREE ENERGY – THE MEASURE OF FEASIBILITY

There was, however, a problem in using entropy change as a measure of feasibility. The entropy changes of system and surroundings had to be computed separately and added. To circumvent the problem, a new function or property was coined called the "Free Energy" (and its variants, "Gibbs Free Energy" and "Helmholtz Free Energy" also called the "Work Function"). For the common isothermal reaction, Free Energy of a system is the sum of the entropies of the system and surroundings multiplied by the absolute temperature (with a negative sign). Thus, Free Energy is another way of expressing Entropy.

The search for the measure of feasibility of reactions ended here with the realisation that the heat liberated goes towards increasing the entropy of the surroundings. The liberated heat has an important role to play in this after all.

2.8. TERMS USED IN THERMODYNAMICS

The science of thermodynamics has evolved from the mathematical treatment of common observations related to heat and work. The terms thus evolved have primarily mathematical definition and to find physical significance have been a somewhat tortuous process. We have seen that Entropy, which is the mathematical ratio of energy per unit absolute temperature, has been found to represent the extent of disorder.

If an object is raised to a higher level, it gains in potential energy. If it is allowed to drop, the potential energy gets quickly converted to kinetic energy until it impacts the hard ground. Immediately after impact all the potential and kinetic energy is lost. Where has all this energy gone? Some is converted into sound, some light (spark), some heat and some ground vibrations. Finally all that energy gets converted into heat or the "Internal Energy" of the affected molecules.

Thus, "Internal Energy" is related to the state of the molecules or atoms – all the energy contained within them – including kinetic energy (vibrations in case of solids and velocity of movements in case of fluids). For real processes – processes realised in practice – most of the changes take place at atmospheric or nearly constant pressure. At constant pressure conditions, it is the "Enthalpy" or "Heat Content" which is more relevant. It is the sum total of the "Internal Energy" and the work it has already performed on the surroundings.

The term "Free Energy" was probably coined originally to signify a part of the energy, which can be easily released. If substantial "Free Energy" is available which can easily leave the system, the transformation is easily feasible. If no such "Free Energy" were available (is zero), the system would not materially change (*i.e.*, it is in thermodynamic equilibrium).

In contrast "Entropy" represents the quantity of bound energy in the system per unit absolute temperature. This bound energy is not normally available for reactions. And "Free

Energy" is obtained by deducting this "Bound Energy, T*S" from the Total Energy which is represented by "Enthalpy" or "Heat Content, H".

2.9. BALANCE OF NATURE – IS IT REALLY ORDERLY?

We see rivers flowing continuously into the sea. We know that with the aid of the sun, seawater evaporates and joins the clouds. Clouds, with the aid of winds (generated largely by differential heating by sun) reach mountain peaks and build up further the ice caps already present. In another stage, sun helps to melt the ice caps, which maintain the flow of water in perennial rivers. The situation appears so well balanced; so much in equilibrium; so orderly. Is it really so?

The balance indicated above, as also the balance of life, balance of nature – are all maintained through the receipt of energy from the sun. Most of that energy received is dissipated to other parts of the universe – adding to the "disorder" or the "free energy" there.

2.10. ENTROPY AND THE BIG BANG

We are destined to be in a world where disorder or Entropy has to be increased every moment. Free Energy has to be dissipated out continuously to areas, which can absorb it. The Universe was apparently formed by a "Big – Bang". The fact that the distant stars and the constellations continuously move away from each other, as also from our world, gives further weight to the thermodynamic principle that the disorder (Entropy) in the universe is constantly increasing.

2.11. THERMODYNAMICS, BILLIARDS TABLE AND COMMON SENSE

Feasibility predictions on the basis of thermodynamic properties can be illustrated by the example of an inclined billiards table with a large number of balls on it. We assume that the balls can move with very minimum friction and very small loss of momentum on impact. Let us also assume that the balls at zero time are in motion uniformly all over the table. By laws of mechanics, it is possible to predict the position of the balls after lapse of different amounts of time. But by common-sense, we know that, after we allow substantial time, the balls will tend to concentrate on the lower area of the inclined table. Predictions of feasibility by thermodynamic laws are of the latter type. And why not, since Thermodynamics is based on common-sense.

Chapter- 3

Basic Concepts

3.1. WHAT IS THERMODYNAMICS?

In simple terms, Thermodynamics deals with transformation of heat into other forms of energy and vice versa. But it has been defined variously as follows:

- Thermodynamics is a science dealing with changes in energies of systems and transformations of energy within a system.
- Thermodynamics is a science that deals with heat and work and those properties of substances that bear a relation to heat and work *(Wylen 1959 & Rock 1969).*
- Thermodynamics concerns itself with the study of energy and the transformations of that energy *(Morrill 1972).*
- Thermodynamics deals with the conversion of thermal energy into work.

Historically, this branch of science arises from the study of heat. The basis of thermodynamics is observation, both experimental and common. Such studies and observations have been made on 'Systems' and 'Surroundings,' which are common concepts, but need to be more precisely defined here.

3.2. WHAT IS A 'SYSTEM'?

A 'system' is that part of physical world, which is under consideration.

This statement is so simple that it appears superfluous. But it needs to be stated to avoid any ambiguity.

The rest of the physical world is called 'surroundings.' System and surroundings together make up the physical world, which we often refer to as the universe. System is separated from surroundings by 'system boundaries'. These boundaries may either be movable or fixed. For example, gas inside an inflated balloon can be considered as a system. The inner surface of the balloon would then be the system boundary. During inflation by blowing or through warming (as when exposed to sun), the boundary moves.

A system may be open, closed or isolated. An 'open' system is one, which can exchange energy and matter with the surroundings.

A 'closed' system cannot exchange matter with the surroundings (exchange of energy may take place).

An 'isolated' system has a definite amount of energy and matter and it cannot exchange them with the surroundings.

In the example of balloon above, as long as we keep blowing in the gas, it is an open system and when we seal the mouth of balloon it becomes a closed system. But it is not an isolated system since the gas inside can get heated or cooled due to the condition of its surroundings. When we carry ice or hot tea in a thermo-flask we try to create an isolated system.

3.2.1. Homogeneity and Heterogeneity

A homogeneous system is completely uniform throughout *(Glasstone 1947)*.

A system is heterogeneous when it is not uniform. Non-uniformity may be of two types; in one case the concentration of species may vary from one point to another; and in the second case the system may consist of two or more 'phases,' which are separated from one another by definite bounding surfaces (three-dimensional boundaries) *(Glasstone 1947)*.

A 'phase' is a chemically homogeneous part of a system, separated by definite bounding surfaces from other similar phases. Within a phase, concentration gradient *i.e.*, variation of concentration from point to point, may sometimes exist, but it would always tend to homogenise itself.

3.3. STATE AND THERMODYNAMIC PROPERTIES

The 'state' of a system is defined by the properties the system exhibits *(Glasstone 1947)*.

If these properties do not 'tend' to change with time, the system is said to be in an 'equilibrium state.'

Here the term 'tend' needs a bit of investigation and inferring, since it is always difficult, and sometimes not possible to measure tendencies. They are often inferred indirectly based on experience. Water level in the reservoir upstream of a dam 'tends' to go down to the level in the river downstream. This tendency we can infer by providing an imaginary passage of flow between the two levels and estimating pressure on an imaginary membrane separating the two waters in the passage.

At 'equilibrium', there is no tendency for any net transfer of matter or energy across the boundary between phases making up a heterogeneous system. In a homogeneous system in equilibrium, there is no tendency for a net transfer of matter or energy within the phase *(Sage 1965 & Gaskell 2003)*. The term 'equilibrium' takes slightly different meaning in other usage. Therefore, for being more explicit, the term 'thermodynamic equilibrium' is used to convey the present meaning more emphatically.

In equilibrium state, the properties tend to be interdependent. It is possible that only a limited number of properties will completely define an equilibrium state of a system. If these properties do not depend upon the path the system has taken to reach that equilibrium state, then they are called 'State Properties.' Mathematically, State Properties are 'Exact Differentials.'

3.3.1. State Properties

If any thermodynamic property 'G' of a system is a single-valued function of certain variables x, y, z, etc., which again are the properties of the system; then G is called a state property of that system. It means that G does not depend upon the path taken to bring the system to that state or condition and depends only on the properties of the system in that state. For example, the state of one mole of an ideal gas is completely defined by defining pressure and temperature, and under these defined conditions, it as a definite specific volume. All the three *i.e.*, pressure, temperature and specific volume of an ideal gas are its state properties.

Mathematically such a function is called an exact differential and the following relationship holds good

If $$G = f(xyz \text{ ....})$$

then $$dG = (\partial G/\partial x)_{y,z} dx + (\partial G/\partial Y)_{x,z} dy + (\partial G/\partial z)_{x,y} dz + \quad \text{.........}$$

For example, a molar volume of an ideal gas is brought from a state (P, V, T) to a state ($P + dP$, $V - dV$, $T + dT$), then whatever way the change is brought about, the following interrelation will hold.

$$PV/T = ((P + dP)\ (V - dV))/(T + dT)$$

Let us assume that at first, temperature is kept constant and the gas is compressed to volume ($V - dV$). Let the increase in pressure be dP_1

By definition,

$$(\partial P/\partial V)_T = dP_1/dV$$

or, $$dP_1 = (\partial P/\partial V)_T\, dV$$

Now if the temperature is increased to $T + dT$ keeping the volume constant at $V - dV$, the pressure of gas will increase again. Let this increase be dP_2.

Again by definition,

$$(\partial P/\partial T)_V = dP_2/dT$$

or, $$dP_2 = (\partial P/\partial T)_V\, dT$$

But according to the gas law $dP_1 + dP_2$ must equal dP.

That is, $$dP = dP_1 + dP_2$$

$$= (\partial P/\partial V)_T\, dV + (\partial P/\partial T)_V\, dT$$

3.3.2. Intensive and Extensive Thermodynamic Properties

A system will have, at any instant, some definite properties. Properties, which are relevant to the thermodynamic study of the system, are called thermodynamic properties. These can be further classified into:

(*i*) Extensive properties,

(*ii*) Intensive properties.

Extensive property of a system is a thermodynamic property, which is dependent upon the quantity of matter in the system (*e.g.*, volume, energy, mass). Extensive properties are additive. Its value for the whole system is the sum of the values of individual parts.

Intensive property is a thermodynamic property, which is independent of the mass of the system. These properties are not additive (*e.g.*, temperature, pressure). The value for the whole systemis not the sum of the values for different parts.

3.4. THERMODYNAMIC EQUILIBRIUM

A system is said to be in Thermodynamic Equilibrium State when its state properties have defined values, which do not tend to change with time *(Khangaonkar 1967).*

It may be noted that though mass is an extensive property, density is an intensive property; volume is an extensive property but 'molar volume' is an intensive property. Here the term 'molar' means molecular weight expressed in weight terms. Thus, a gram-mole of oxygen refers to 32 grams of oxygen. By analogy, a gram-atom of carbon refers to 12 grams of carbon. Molar properties come in handy in treatment of ideal gases for ease of comparison. For example, molar volumes of all ideal gases are same at a particular temperature and pressure.

3.5. WORK - DIFFERENT FORMS

We say that 'work is done' when a body moves under the influence of a force.

When the point of application moves in the direction in which the force acts, work is said to be done by the force (or 'the force has done the work').

If the point of application moves in the direction opposite to that of the force, work is said to be done against the force (or 'work has been done on the force').

Work is generally denoted by the symbol 'W'. If a system expands against the incumbent pressure of the surroundings, the system is said to perform work on the surroundings and W is assumed to be positive.

If the system contracts under the influence of the pressure of the surroundings then 'W' is assumed to be negative. We have given here examples of only mechanical work. In fact, work can be of chemical type also. In a chemical reaction, when a molecule is broken, work is performed to break the bond, which had been holding the atoms together in the molecule. On the other hand, work is done by the atoms, when joining together to form new bonds and thereby a new molecule is formed. Work can also be electrochemical when the applied electrical potential (e.m.f.), forces the ions to break the electrovalent bond and then move them towards a particular electrode. Work can also be electrical work, when electrons move in conductors, overcoming its resistance and applied potential.

3.6. BRANCHES OF THERMODYNAMICS

As stated earlier, Thermodynamics deals with the conversion of thermal energy into work. The work may be mechanical, or of any other type.

Mechanical work manifests itself into displacement of a body under mechanical force such as accomplished in heat engines. For example, steam locomotives perform mechanical work. Study of such processes has lead to one particular branch of thermodynamics, which is commonly referred to as 'Thermodynamics Applied to Heat Engines' or 'Mechanical Engineering Thermodynamics' or simply 'Engineering Thermodynamics.'

However, as mentioned earlier, heat evolution or absorption is also involved when substances react chemically to form new molecules. In forming these new molecules, the atoms in the old molecules have worked against the forces binding it and in the process have used some energy available in the vicinity – releasing some of it when it gets bound into the new molecule. Chemical work is required to be performed to make these new molecules revert back to the reactants. Science of study of energy changes in such a process is referred to as 'Chemical Thermodynamics.'

Science of study of a process where a chemical change generates electricity or vice versa as in the charging of an electro-chemical cell, when the imposed electro-motive force pushes the electrically-charged ions to a higher energy state, is referred to as 'Electro-chemical Thermodynamics.'

Similarly, Metallurgical Thermodynamics is that part of thermodynamics, which deals with energy changes in reactions involving metals and their compounds. Many look upon it as the "High-Temperature Chemical Thermodynamics."

There are many common concepts between Chemical and Metallurgical Thermodynamics. They differ in the ranges of temperature used (high temperatures in metallurgical processes and relatively lower temperatures in chemical processes) and also in the ranges of pressure applied (close to atmospheric and up to a few bars in metallurgical processes, while chemical processes very often use hundreds of atmospheres of pressure).

Matters are made up of small particles such as **molecules** and **atoms**. Thermodynamic laws have been postulated and inferred without looking into the micro-properties or micro-states within the systems. A branch of thermodynamics has evolved, which tries to interpret thermodynamic properties based on the properties of micro constituent of the system. This branch is called the "Statistical Thermodynamics." An offshoot is the "Nuclear Thermodynamics", where matter is treated as another form of energy and role of atomic and sub-atomic particle forms are studied in determining thermodynamic properties.

In contrast, the science dealing with the basic concepts, the thermodynamic laws and their understanding is termed as the "Classical Thermodynamics."

Chapter- 4

Zeroeth Law, First Law and Thermochemistry

4.1. CONCEPT OF TEMPERATURE

We have developed the concept of temperature by experience. It is not possible to quantitatively define temperature but we can describe it as a level of heat just like the level of liquid in a container. Thus, heat flows from a body at higher temperature to a body at lower temperature when brought in thermal contact, just like liquid will flow from a container with higher level to a container with lower level when connected with a tube.

When we study heat energy, we necessarily need to study temperature, since we sense heat only indirectly by its effect on temperature. It is true that under some conditions (*e.g.*, isothermal melting of ice), heat transfer does not lead to temperature change but, in general, study of temperature, and the changes therein, are important for heat measurement.

4.1.1. Zeroeth Law of Thermodynamics

When two bodies have thermal equilibrium (equality of temperature) with a third body, they in turn, have thermal equilibrium (equality of temperature) with each other *(Rock 1969)*.

This fact appears very obvious as it is based on our day-to-day experience. Since this fact is not derivable from other laws, and since it logically precedes the first and the second laws of thermodynamics, it has been called the 'Zeroeth Law of Thermodynamics.'

4.1.2. Pure Substance

A pure substance is one that has a homogeneous and invariable chemical composition *(Wylen 1959)*.

4.1.3. Independent Properties of a Pure Substance

A pure substance in the absence of motion, gravity, surface effects, electricity and magnetism, has three intensive properties only two of which are independent, *viz.*, pressure, temperature and concentration (conclusion based on experimental or day-to-day observation). The two independent intensive properties are often referred to as the two "degrees of freedom." For example, if we keep water vapour in an evacuated chamber – say, above its critical temperature - we may be able to change the conditions to a new temperature and pressure. But by doing so, the concentration (vapour density) gets fixed and it does not remain our choice. If we want the water vapour to have a particular concentration or vapour density at a particular pressure (vapour pressure), that would be possible only at a particular temperature, and we cannot enforce a temperature of our choice.

We have used the term 'critical temperature.' Gases can normally be liquefied by applying additional pressure, but not above its 'critical temperature.' Critical temperature is a property of a pure gas. Applying pressure alone above the critical temperature cannot liquefy a gas.

We cannot think of liquefying oxygen, hydrogen, nitrogen, etc., by compressing only without refrigeration as critical temperatures are well below –150°C. Compressed CO_2 gas in a pressurised cylinder exists in gaseous form. But on bleeding out the gas, expansion causes it to cool below its critical temperature of about – 40°C and gas can be seen to come out like a smoke, carrying with it small particles of solid CO_2.

4.2. PHASE RULE

Extended observations on the above lines have led to the Phase Rule, which is best stated mathematically as (at equilibrium)

$$P + F = C + 2$$

where,
P = Number of phases present
F = Number of degrees of freedom
C = Number of constituents

In the example given above – example of a pure substance (H_2O) – value of C is 1 and of P is 1 (water vapour *i.e.* gas phase); therefore, we find that the value of F is 2. For the same pure substance (H_2O), if we select a condition where all three phases, solid (ice), liquid (water) and gas (water vapour), co-exist, this would be possible only at one possible combination of temperature, pressure (vapour pressure) and concentration (the densities of the three phases). Any change in one of the concentrations would cause disappearance of one of the three phases. Thus, there is no degree of freedom available and hence $F = 0$.

Number of degrees of freedom F is not easy to conceptualise but the effect is easily felt when restrictions are imposed on F. The above example is an example of restricted degrees of freedom.

4.3. HEAT

Heat is that form of energy, which is transferred by virtue of a temperature difference or temperature gradient.

4.3.1. Comparison of changes of Heat and Work in a System

(*a*) Both are transient phenomena. Systems never possess change of heat or work but either or both may occur when a system undergoes a change of state.

(*b*) Both are path functions and inexact differentials (*i.e.*, $y \neq (\partial y/\partial x)_t\, dx + (\partial y/\partial t)_x\, dt$ – please refer earlier sections).

However, there are important differences between the two, which is the foundation of the Second Law, about which we would discuss at an appropriate place.

4.3.2. Energy

It is a property by possessing which the system gets the capacity to do work. It is also defined as 'any property, which can be produced from or converted into work (*e.g.*, heat).'

Energy is generally denoted by symbol 'Q'. Q is positive by convention, if energy is transferred to, *i.e.*, taken up by the system.

When a system evolves heat, *i.e.*, it loses energy to the surroundings, Q is then negative.

4.4. THE FIRST LAW OF THERMODYNAMICS

Mathematically, the first law is expressed as

$$dE = dQ - dW^*$$

* Many do not consider the notations dQ and dW to be proper since neither Q nor W are exact differentials.

dE refers to the increase in internal energy of the system.

dQ refers to the energy supplied to the system.

dW refers to the work done by the system.

It is basically a statement of law of conservation of energy. For example, "In a frictionless kinetic system of interacting rigid elastic bodies, kinetic energy is conserved" or

"It is impossible to do work without expending an equivalent amount of energy" or

"Perpetual motion of first kind is impossible".

4.4.1. What is Internal Energy and Heat Capacity?

Consider a system which does not do work, *e.g.*, 1 mole of an ideal gas at constant volume. When energy is supplied to it in the form of heat, its temperature rises. There will be an increase in internal energy, or the energy contained within, corresponding to the heat put in.

This obviously manifests itself in a temperature rise (consequently when this gas loses energy by doing work on the surroundings, such as by adiabatic expansion, the temperature falls). Whatever this internal energy is, it varies with respect to the temperature change in the system, *i.e.*,

$$dE \propto dT$$

or

$$dE = C_v\, dT$$

or

$$C_v = (dE/dT)_v$$

Where, C_v is a constant of proportionality. It is called 'heat capacity at constant volume.' Please note that the above assumes absence of any phase change, such as **liquefaction**.

4.4.2. Absolute Value of Internal Energy?

We have mathematically defined internal energy only by stating the change therein, i.e. dE. It is impossible to determine the absolute value of E. A suitable reference point may be assumed with zero E for ease of calculations.

4.4.3. Enthalpy and Heat Capacity

Correspondingly, we can have a term called 'heat capacity at constant pressure.' When a system of say, 1 mole of ideal gas, is supplied with energy in the form of heat dQ, and the super-incumbent pressure is maintained constant, the gas will expand. Still, after expansion, the temperature will be seen to rise. If this rise is dT, then we can define heat capacity at constant pressure C_p as

$$dQ = C_p\, dT$$

or,

$$C_p = (dQ/dT)_p$$

Now this heat dQ has been taken up by the system and it has resulted in

(*i*) Increase in the internal energy of the system (temperature has risen), and

(*ii*) Work done by the system (gas has expanded against external pressure).

Thus, from First Law, we have

$$dE = dQ - dW$$

$$= dQ - PdV$$

or

$$dQ = dE + PdV$$

$$= d(E + PV) \quad \text{since } P \text{ is constant}$$

For study of constant pressure processes, it is convenient to define a term called 'Enthalpy' or 'Heat Content', H, equal to $E + PV$. Like E, it is impossible to know the absolute value of H for a system. But, for convenience, H for pure elements at atmospheric pressure and 25 °C (298 °K) is taken to be zero.

4.4.4. Expansion of An Ideal Gas

In the above example, when the system of ideal gas expands under constant pressure

$$dQ = dE + PdV$$

or

$$C_p dT = C_v dT + PdV$$

or

$$C_p = C_v + PdV/dT = C_v + d(PV)/dT = C_v + d(RT)/dT$$

Hence

$$C_p = C_v + R$$

Thus for an ideal gas

$$C_p - C_v = R$$

4.4.5. Reversible Process

It is a process in which the system is never more than infinitesimally removed from a state of thermodynamic equilibrium (and to which, therefore, the equation of state is applicable – *Rock 1969*).

To effect a finite change, such a process would take infinite time. Thus, all naturally-occurring processes, which take finite time, are irreversible.

Also we can state that a natural process, which satisfies the condition of thermodynamic reversibility, is in thermodynamic equilibrium.

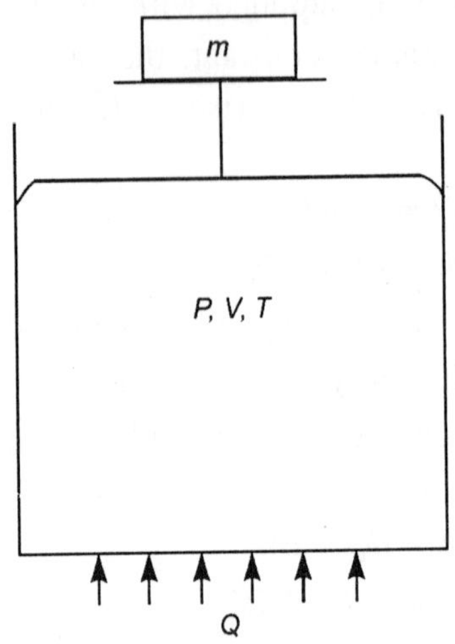

Figure 4.1 Illustration of reversible process

4.4.6. Reversible Isothermal Expansion of an Ideal Gas

Since this process is isothermal, hence $dT = 0$

i.e., $dE = 0$

Hence, $dQ - dW = 0$

And $dQ = dW$

Let us consider one mole of an ideal gas contained in a cylinder with a frictionless piston.

The pressure P of gas is balanced by atmospheric pressure and the combined weight of the piston and the weights kept on it.

If the weight on the piston is slowly decreased such that the pressure exerted by the piston is always infinitesimally less than the pressure of the gas, then the piston will start moving up very slowly.

If, simultaneously, we supply heat to the gas in the cylinder to maintain its temperature at T then we have a case of isothermal reversible expansion. If we can imagine that the weight on piston is made up of fine sand and we imagine that the pressure is released by removing one sand particle at a time then such a process will very closely represent a reversible expansion.

If the expansion in volume is dV, the distance the piston moves is dx and the area of the piston is A, then

$$dV = Adx$$

Work done by the gas = Force * distance

$$= (P.A) * dx$$

$$= P.Adx$$

$$= PdV$$

That is $dQ = PdV$

Please note that pressure has been changing (as we remove sand particles) in this example.

Thus, if a finite amount of energy Q is supplied for a finite expansion of volume from V_1 to V_2 (consequently pressure decreases from P_1 to P_2)

then, $$Q = \int_{V_1}^{V_2} PdV$$

$$= \int_{V_1}^{V_2} \frac{RTdV}{V}$$

$$= RT[\ln V]_{V_1}^{V_2}$$

That is $Q = RT \ln (V_2/V_1)$

Also, $P_1V_1 = P_2V_2$

Hence, $V_2/V_1 = P_1/P_2$

And $Q = RT \ln (P_1/P_2)$ (isothermal reversible)

We would be using this relationship later on.

4.4.7. Adiabatic Reversible Expansion

If the expansion is adiabatic $dQ = 0$

Therefore, $dE = -dW = -PdV$

or $C_v dT = -PdV$

$$= -(RT/V)\,dV$$

or $C_v\,dT/T + R\,dV/V = 0$

or $(C_p - C_v)\,dV/V + C_v\,dT/T = 0$

or $(\gamma - 1)\,dV/V + dT/T = 0$ $\quad (\gamma = C_p/C_v)$

or $$(\gamma - 1) \ln V + \ln T = \text{Constant}$$

or $$V^{(\gamma-1)} T = \text{Constant}$$

As $$T = PV/R, \text{ then}$$

$$PV^{\gamma}/R = \text{Constant}$$

or $$PV^{\gamma} = \text{Constant}$$

Since $\gamma > 1$, therefore, for an adiabatic process, pressure drop produces a lesser volume increase than in isothermal process.

4.5. APPLICATIONS OF THE FIRST LAW — THERMOCHEMISTRY

The first law is a restatement of law of conservation of energy, and therefore finds application wherever energy conservation is in question. But while dealing with energy conservation in chemical reactions, the application of first law has led to the evolution of a branch of science called 'Thermochemistry'.

Thermochemistry is that branch of science which is concerned with heat changes associated with chemical reactions. There are several important applications of first law. Two of these examples relate to the following:

1. Hess's law of heat summation
2. Calculation of maximum theoretical (adiabatic) flame temperature in a combustion system.

The following Rule of Lavoisier and Laplace is really a corollary of the Hess's Law.

4.5.1. Rule of Lavoisier and Laplace

The quantity of heat, which must be supplied to decompose a compound into its elements, is equal to the heat evolved when the compound forms from its elements *(Glasstone, 1947)*.

4.5.2. Hess's Law of Heat Summation

The resultant heat change at constant pressure or constant volume in a given chemical reaction is the same whether it takes place in one or several stages *(Glasstone, 1947)*.

Thus, thermo-chemical equations can be added and subtracted just like algebraic equations.

4.5.3. Heat of Formation (of a Compound)

It is the increase of heat content ΔH when 1 mole of a substance is formed from its elements at a given temperature and pressure *(Glasstone, 1947)*.

For example, it is given that, for

$$CO(g) + 1/2\ O_2(g) = CO_2(g)$$

$$\Delta H^0_{298} = -\ 67.65 \text{ kCal per mole}$$

From Standard Tables it is seen that the standard enthalpies of formation of CO and CO_2 are

$$\Delta H^0_{298} = -\ 26.42 \text{ and } -\ 94.05 \text{ kCal/mole respectively.}$$

Do these data violate Hess's law *(Sage 1965; Gaskell* 2003)?

When CO_2 is formed in one step from C and O_2

$$\Delta H^0_{298} = -\ 94.05 \text{ kCal/mole}$$

i.e., $$C + O_2(g) = CO_2(g)$$

$$\Delta H^0_{298} = -\ 94.05 \text{ kCal/mole}$$

When it is formed in two steps

$$C(s) + 1/2\ O_2(g) = CO(g)$$

$$\Delta H^0_{298} = -\ 26.42 \text{ kCal/mole}$$

$$CO(g) + 1/2\ O_2\ (g) = CO_2\ (g)$$

$$\Delta H^0_{298} = -\ 67.65 \text{ kCal/mole}$$

Adding we get

$$C(s) + O_2(g) = CO_2(g)$$

$$\Delta H^0_{298} = -\ 94.07 \text{ kCal/mole}$$

This value closely agrees with the value of 94.05 obtained by direct measurement and hence vindicates Hess's Law.

4.5.4. Heat of Combustion

It is the heat change accompanying the complete combustion of 1 mole of a compound, at a given temperature and one atmosphere pressure.

4.5.5. Theoretical Flame Temperature

When a gaseous fuel burns under conditions that the flame produced is short and non-luminous, then the heat evolved during oxidation of fuel is almost completely taken up by the products of combustion.

From the knowledge of the heats of combustion and the heat capacities of the combustion products and assuming the combustion within flame to be adiabatic, the temperature of the combustion products can be calculated. This would give the theoretical maximum flame temperature, which is also called the **adiabatic flame temperature**.

Example

Heat of combustion of methane is –212.80 kCal at 25 °C; the difference between heat contents of liquid water and water vapour at 1 atmosphere pressure at 25 °C is 10.52 kCal. Given,

$$C_{p\,CO2(g)} = 10.55 + 2.16 * 10^{-3} * T - 2.04 * 10^5 * T^{-2} \text{ Cal/deg.-mole}$$

$$C_{p\,H2O(g)} = 7.17 + 2.56 * 10^{-3} * T + 0.08 * 10^5 * T^{-2} \text{ Cal/deg.-mole}$$

$$C_{p\,N2(g)} = 6.66 + 1.02 * 10^{-3} * T \quad \text{Cal/deg.-mole}$$

Let us calculate the adiabatic flame temperature for the combustion of CH_4 in air (20% O_2 and 80% N_2).

We have

$$CH_4(g) + 2O_2(g) = CO_2(g) + 2H_2O\ (l) \qquad ...(1)$$

$$\Delta H^0_{298} = -\ 212.80 \text{ kCal}$$

By taking out the latent heat of condensation of water vapour we get the net heat of combustion of CH_4 (during combustion water vapour does not condense)

i.e.,

$$2H_2O(l) = 2H_2O(g) \qquad ...(2)$$

$$\Delta H^0_{298} = 2 * 10.52 \text{ kCal}$$

$$= 21.04 \text{ kCal}$$

$$(1) - (2) \equiv CH_4(g) + 2O_2(g) = CO_2(g) + 2H_2O(g)$$

$$\Delta H^0_{298} = -\ 191.76 \text{ kCal}$$

This heat is used to raise the temperature of combustion products, *i.e.*, $CO_2(g)$ and $H_2O(g)$, and also the N_2 associated with air required for combustion.

Quantity of nitrogen associated is 8 moles

Let the final temperature be T_2 (°K), then, we have

$$\Delta H = \int_{298}^{T2} C_{P_{CO_2(g)}} dT + \int_{298}^{T2} C_{P_{H_2O(g)}} dT + \int_{298}^{T2} C_{P_{N_2(g)}} dT$$

$$= [10.55 * T + 2.16 * 10^{-3} * T^2/2 + 2.04 * 10^5 * T^{-1}]_{298}^{T2}$$

$$+ 2 * [7.17 * T + 2.56 * 10^{-3} * T^2/2 - 0.08 * 10^5 * T^{-1}]_{298}^{T2}$$

$$+ 8 * [6.66 * T + 1.02 * 10^{-3} * T^2/2]_{298}^{T2}$$

$$= [(10.55 + 2 * 7.17 + 8 * 6.66) * T + (1.08 + 2.56$$

$$+ 4.08) * 10^{-3} * T^2 + (2.04 - 0.16) * 10^5 * T^{-1}]_{298}^{T2}$$

$$= [78.17 * T + 7.72 * 10^{-3} * T^2 + 1.88 * 10^5 * T^{-1}]_{298}^{T2}$$

Thus,

$191,760 = 78.17(T_2 - 298) + 7.72 * 10^{-3} (T_2^2 - (298)^2] + 1.88 * 10^5 [1/T_2 - 1/298]$

Solving by the method of successive approximation, we get

$T_2 \approx 2250°K$ or ≈ 1980 °C

There are several reasons why the results obtained thus are higher than the experimental flame temperatures. Some of these are listed below.

1. It is unlikely that the reaction can be carried out under such conditions that the process is adiabatic (*i.e.*, no heat is lost to the surroundings). However, when there is adequate premixing of fuel and air so that a short non-luminous flame is obtained the combustion is expected to be very nearly adiabatic and the flame temperature is high.
2. In practice, excess air must be used to effect complete combustion of fuel. The extra oxygen and the nitrogen accompanying it will pick up some heat during combustion, thus reducing the temperature.
3. At the temperature of combustion, there is appreciable dissociation of water vapour into hydrogen and oxygen, or hydrogen and hydroxyl and of CO_2 into CO and O_2.
4. Combustion process is usually a more complex chemical reaction. Often these reactions lead to the temporary formation of solid particles, which incandesce and dissipate heat by radiation.

None the less, by maintaining condition for efficient combustion, flame temperature may reach close to the theoretically calculated temperature.

4.5.6. Kirchoff's Relationship

In the last example of calculation of theoretical flame temperature, we have assumed that the combustion has taken place at room temperature and the heat liberated has heated the reaction products to the flame temperature. We have integrated $Cp\ dT$ values of the products up to the final temperature to calculate the heat absorbed in heating the product gas from room temperature to the final temperature.

Although combustion does not actually take place at room temperature, Hess's Law gives us a justification of making calculations in the way we have done it here. This way of making calculation is also convenient for us since heat of combustion values have often been determined, tabulated and reported in such a way that the room temperature values can be easily retrieved.

Many times we need to have the combustion or reaction data at a temperature at which the data are not normally reported in literature. By using Hess's Law and change in heat capacity values, the heat of reaction at temperature of interest can be calculated.

From the Hess's Law, or the First Law of Thermodynamics the following relationship between the enthalpy change and the change in heat capacity in a reaction can be derived.

$$d(\Delta H) = \Delta C_p dT$$

This relationship is valid for reaction of the type considered in the last example, *i.e.*,

$$CH_4(g) + 2O_2(g) = CO_2(g) + 2H_2O(g)$$

$$\Delta H^0_{298} = 191.76 \text{ Kcal}$$

And in this derivation, it has also been assumed that the reaction is carried out at constant pressure.

By a knowledge of ΔCp value (which is often a temperature dependent property and generally of the form $A + BT + CT^{-2}$) the above expression can be integrated between 298°K and the temperature of interest, to obtain the enthalpy change at the desired temperature.

For constant volume processes the relationship is modified into

$$d(\Delta E) = \Delta C_v dT$$

These two relationships are called Kirchoff's relationships and are, in fact, a direct application of the First Law of Thermodynamics.

4.5.7. Theoretical Explosion Temperature and Pressure

Theoretical explosion temperature of explosive mixtures can be calculated in the same manner as flame temperature except for the following:

(*i*) Using ΔE values in place of ΔH values

(*ii*) Using C_v in place of C_p

The maximum pressure can be estimated assuming ideal gas laws are applicable.

We have given in an earlier section (2.1) the example of hydrogen and oxygen mixture at room temperature and simultaneously have cautioned against making such a mixture. Such a mixture is, in fact, one of the possible 'explosive mixtures' of gases. The theoretical explosion temperature and pressure of this mixture can give a comparative idea of the enormity of damage this mixture can inflict.

Example

Let us calculate the temperature and pressure (starting with 1 atmosphere pressure of mixture) obtained in such an explosion using the following data:

$$H_2(g) + ½\, O_2(g) = H_2O(g)$$

$$\Delta H^\circ_{298} = -57,560 \text{ Cal/mole, and}$$

$$C_{v\,H2O(g)} = 5.18 + 2.56 * 10^{-3}*T + 0.08 * 10^5 * T^{-2} \text{ Cal/deg-mole}$$

Let the adiabatic explosion temperature be T

or $$5.18\,(T - 298) + 1.28 * 10^{-3}(T^2 - 298^2) + 8000((1/298) - (1/T)) = 57,560$$

or $$57,560 = 1.28 * 10^{-3}T^2 - (8000/T) + 5.18T - (5.18 * 298 + 1.28 * 0.298 * 298 - 8000/298)$$

Solving by the method of successive approximation we get

$$T = 5060°K \text{ or } 4787\ °C$$

And at this resultant temperature the pressure developed would be 5060/(298*1.5) = 11.32 atmospheres. However, complete attainment of reaction to reach 4787 °C temperature is open to question because of uncertain stability of H_2O molecules above about 3150 °C. But even if temperature was only 3150 °C; the pressure developed would be 8.6 atmospheres.

Explosion is a much more complicated phenomenon than what has been presented here. The above calculated temperature (3150 °C) and pressure (8.6 atmosphere) values should be taken as the upper limits in explosion of this mixture. In a few sets of controlled experiments on explosion of various mixtures of hydrogen and oxygen, as reported by *Lewis & Elbe (1987)*, the maximum observed pressure was a little over 8.2 atmospheres and the maximum observed temperature was a little over 2460 °C.

4.5.8. Heat Capacities at Constant Pressure and Constant Volume

We have defined heat capacities for constant volume and constant pressure processes as follows

At constant volume

$$C_v = dE/dT$$

At constant pressure

$$C_p = dH/dT$$

C_p and C_v are interrelated by the work of expansion

$$C_p - C_v = P * dV/dT$$

For an ideal gas, the right hand term would equal the gas constant R. In CGS unit, the value of R is 1.987 calories/deg.K-gm-mole.

From the point of view of metallurgists, C_p is of vital importance as most of the metallurgical processes take place at constant pressure. On the other hand, many chemical syntheses are carried out at high pressures in confined reactors at constant volume (for example, synthesising ammonia from nitrogen and hydrogen gases). Under these conditions, it is the C_v which should be considered.

In the laboratory, C_p values are determined by measuring the change in heat content between room temperature (or calorimeter temperature - calorimeter is the measuring apparatus for heat) and a number of higher temperatures. The calorimeter commonly used for this purpose is the isothermal ('isoperibol' is a more appropriate term) room temperature calorimeter, which uses the drop method. The substance is heated to the desired temperature and is dropped into the calorimeter and the heat effect is measured. This technique is commonly referred to as the 'mixture method' in calorimetry.

If there are no phase changes in this temperature range then the heat effect measured by the calorimeter is related to the heat capacity thus

$$\Delta H_{298}^{T} = \int_{298}^{T} \Delta C_p \, dT$$

Attempts were made to select an expression of C_p in term of T so that most of the results can be fitted in a single expression, which would enable easy tabulation of data. An expression of the type

$$C_p = a + bT + CT^2$$

has been found satisfactory in most cases. However, Kelley found that an expression of the type

$$C_p = a + bT + cT^{-2}$$

fits the results even better. In this case, the experimentally-measured enthalpy should fit in the following curve

$$\Delta H_{298}^{T} = A + BT + CT^2 + \mathrm{D}\mathrm{T}^{-1}$$

where, $B = a$, $C = b/2$ and $D = -c$

A, B, C & D are evaluated from four or more points on $\Delta H \rightarrow T$ curve.

4.5.9. Calorimeter

As mentioned earlier, we do observe that a burning candle can make water in a test tube boil in minutes while on a bucketful of water it cannot do so even in hours. Water in the bucket absorbs much more heat obviously than water in test tube without corresponding rise in temperature. So, while we have developed the concept of temperature through our senses, the same temperature is not a measure of the heat contained in or transferred to a system. For the measurement of heat we normally take the help of a calorimeter.

The word 'calorie' is another name for heat. By usage 'calorie' denotes that amount of heat which one gram of water takes up (at constant pressure) in raising its temperature by 1°C.

As mentioned in the last section, the heat capacities of solid substances are measured using mixture method in a calorimeter. Special calorimeters are required for specific determinations. Isothermal ice calorimeter is used for measuring heat effect at 0 °C. ***Iso-peribol calorimeter*** is the most common for measuring heat effects at room temperatures - mostly of chemical reactions in liquid solutions - or where the product is a liquid solution. The term Iso-peribol implies ***isothermal surroundings***. When this isothermal surrounding is maintained using predominantly water medium, it is referred to as submarine type calorimeter. Then there are ***adiabatic calorimeters***, ***constant heat flux calorimeters***, ***calorimeters*** measuring heat effects through ***thermal analysis technique***, etc.

Chapter- 5

Second Law of Thermodynamics

5.1. THE SECOND LAW – DIFFERENT STATEMENTS

The First Law presents the interchangeability of heat and work. The second law states the limitations on heat when it is converted into work.

The second law of Thermodynamics has been stated in many ways, some of which are:

(*i*) It is impossible to devise a machine, which would work in cycles and produce no effect other than to extract heat from a source and convert it completely into work.

(*ii*) It is not possible to convert heat into work by a constant temperature cycle.

(*iii*) Heat cannot, by itself, flow from a cooler to a hotter body.

(*iv*) Carnot's theorem/principle: All reversible heat engines operating between two given temperatures have the same efficiency.

5.1.1. Are These Statements Really Same?

To understand these statements, let us consider a simple system of one mole of an ideal gas in a cylinder with a piston as illustrated in Fig. 4.1. Let this gas be at temperature T_1, pressure P_1 and volume V_1.

Let the walls of the cylinder be thermally insulating and its base be thermally conducting which is kept in contact with a heat reservoir at temperature T_1. Let the weight of the piston and the external pressure and load on it be such that it just balances the pressure of the gas. In such a situation, there will not be any net heat flow between gas and the heat source.

Now if the weight on the piston is very slowly reduced so that the gas expands reversibly, the temperature of gas will tend to fall as it is performing work against the external pressure. The gas will therefore pick up heat from the heat reservoir to maintain its temperature. Suppose the gas expands to a volume of V_2 so that the pressure is P_2.

Let Q_1 be the heat taken up by the gas. Then according to the first law, Q_1 should equal the work done by the system.

$$Q_1 = \int_{V_1}^{V_2} PdV = \int_{V_1}^{V_2} RT_1 * dV/V$$

$$= RT_1 \ln (V_2/V_1)$$

also $$= RT_1 \ln (P_1/P_2)$$

Here heat has been completely converted into work but the system is not in the same state as it was to start with. The system can be brought to its initial state by reversibly compressing the gas to a pressure of P_1. The gas will give out heat to the heat reservoir so that its temperature is maintained at T_1. By the time the pressure P_1 is reached, an equal amount of work, as was performed by the gas during expansion from P_1 to P_2, will have to be done on it. Consequently, the gas will return an equal amount of heat (Q_1) to the heat source. Hence, after such a system is made to perform in a cycle, the net effect is that no heat is taken up or given out by the system and no work is done by or on the system. This type of experience is compatible with statement (*ii*) of the second law, which says that it is not possible to convert heat into work by a constant temperature cycle.

5.1.2. Efficiencies of Reversible Heat Engines

Most of the statements of second law pre-suppose the existence of a heat engine or a machine that converts heat into work. From the example cited above, it is clear that if we do conceive of a heat engine it must be working between at least two different temperatures. Before we visualise such a machine, let us assume that such a machine does exist and it can act in cycles in a reversible manner so that we can examine statement (*iv*) of second law.

Let there be two heat sources as represented in Fig. 5.1, one at temperature T_1 and the other at a lower temperature T_2. Let there be two heat engines I and II working between the two temperatures in cycles in a reversible manner. To start with, let us assume to the contrary that the two heat engines have different efficiencies. Suppose each of them take up heat Q_1 from the higher temperature source T_1 but perform different amounts of work W_1 and W_2 before discharging the balance amount of heat Q_2 and Q_2' to the other heat reservoir at temperature T_2.

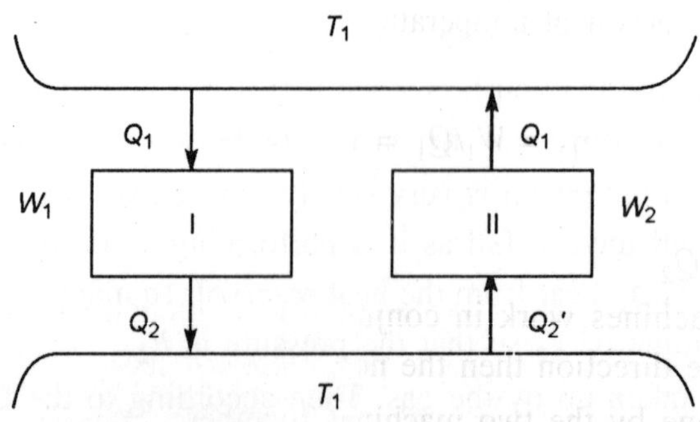

Figure 5.1 Two reversible heat engines working in conjunction

Therefore, from the first law we have,

$$W_1 = Q_1 - Q_2$$

and $$W_2 = Q_1 - Q_2'$$

and the efficiencies of the heat engines are

$$\eta_1 = W_1/Q_1 = (Q_1 - Q_2)/Q_1$$

$$= 1 - Q_2/Q_1$$

and $$\eta_2 = W_2/Q_1$$

$$= (Q_1 - Q_2')/Q_1 = 1 - Q_2'/Q_1$$

Let $W_2 < W_1$; then $\eta_2 < \eta_1$

The engines being reversible, their direction of working can be reversed by only infinitesimal change in the constraints. Thus, engine II will work taking up Q_2' amount of heat from reservoir at T_2 and discharging Q_1 amount of heat to the reservoir at T_1 provided W_2 work is done on the machine reversibly by an external agency.

If the two machines work in conjunction in the manner shown in the diagram then the net effect of the two machines are

(*i*) $Q_1' - Q_2$ amount of heat are taken up from the reservoir at temperature T_2

(*ii*) $W_1 - W_2$ amount of work is done by the machine $(= Q_2' - Q_2)$

(*iii*) No heat is given out to the reservoir at temperature T_1.

Thus, the machines have in one cycle taken up $Q_2' - Q_2$ amount of heat and has completely converted into work $(W_1 - W_2)$ without bringing out any other change. This violates the statement (*i*) of second law as long as $\eta_1 \neq \eta_2$.

Thus, we see that statement (*iv*) follows directly from statement (*i*).

Now to compare statements (*iii*) and (*iv*) let us assume that in one cycle the machine II performs the same amount of work as machine I.

Let machine II take up Q_1' amount of heat from reservoir at temperature T_1 and discharge Q_2'' amount of heat to reservoir at temperature T_2.

Then,

$$\eta_2 = W_1/Q_1 = 1 - Q_2''/Q_1'$$

hence, $Q_1' > Q_1$

and $Q_2'' > Q_2' > Q_2$

Now if the two machines work in conjunction so that work done by machine I drives machine II in the reverse direction then the net effect will be

(*i*) No work is done by the two machines together.

(*ii*) An amount of heat $Q_2'' - Q_2$ is extracted from reservoir at temperature T_2.

(*iii*) An amount of heat $Q_1' - Q_1$ is discharged into the reservoir at higher temperature T_1 (This amount would obviously equal $Q_2'' - Q_2$).

Thus, the assembly of machines would produce no other effect than to transfer heat from a cooler body (T_2) to a hotter body (T_1) thereby violating statement (*iii*) of the second law. For statement (*iii*) to be valid η_1 must equal η_2, which is same as statement (*iv*). We, therefore, see that statement (*iv*) is also derivable from statement (*iii*).

5.2. CARNOT'S CYCLE

Since all reversible heat engines working between the same two temperatures will have the same efficiencies, we can conclude that their efficiencies depend only upon the two temperatures between which they work. For further thermodynamic consideration it is, therefore, sufficient that we consider that type of reversible machine, which will lend itself to simple thermodynamic treatment. A machine employing Carnot's cycle is of such a type.

Let us consider one mole of an ideal gas contained in a cylinder with a frictionless piston. Statement (*ii*) of second law requires that if this assembly is to be used as a reversible heat engine, it must work at least between two temperatures. Let those temperatures be T_1 (higher) and T_2 (Lower). It is obvious that such an engine cannot work only in adiabatic cycles. An adiabatic cycle will not pick up any energy and therefore cannot give out energy in the form of work.

The system considered must be effecting reversible heat exchanges at the two temperatures. Thus, we must have heat reservoirs at the two temperatures.

In order that the heat exchanges between the gas and the reservoirs take place reversibly it is essential that the temperature of the gas is made equal to that of the particular reservoir before they are brought into thermal contact. This change in temperature must be effected reversibly and without recourse to any other heat reservoir. Such a change can be effected by adiabatic and reversible compression or expansion of the gas.

The pressure vs. volume plots for the ideal gas are hyperbolae at different temperatures. Let us consider the plots at temperatures T_1 and T_2, the temperatures of our interest, as in Fig. 5.2.

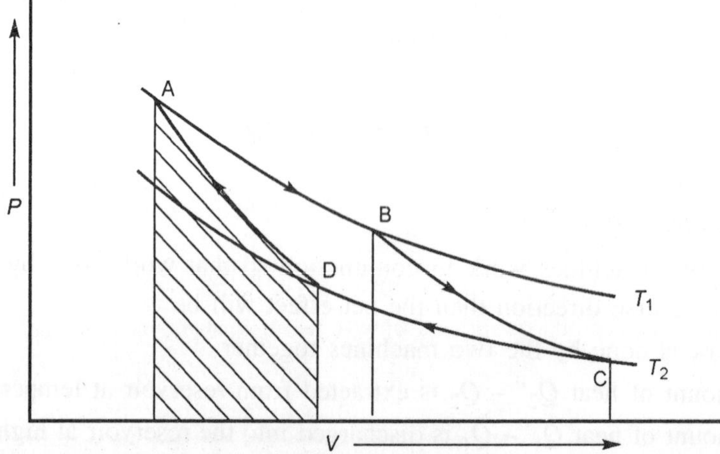

Figure 5.2 Carnot's cycle

Let us also assume that the gas in the cylinder is at point A to start with, *i.e.*, it is at a temperature T_1, pressure P_A and volume V_A. If it is allowed to expand adiabatically and reversibly so that the final temperature is T_2, then the state of gas will be represented by point D, *i.e.*, the gas now has temperature T_2, pressure P_D and volume V_D (the path in this instance is opposite to the arrow mark). The line AD represents the path the gas has followed during adiabatic expansion from A to D. The AD curve is in fact the plot of PV^{γ} = Constant passing through point "A". The work done by the gas is represented by the area under the curve AD (shaded region). If the gas is again adiabatically compressed from D to A (towards the arrow mark), an equal amount of work will have to be done on the gas and the gas would return to its original state.

Let the gas at A be brought in thermal contact with the heat reservoir at T_1 and be allowed to expand reversibly and isothermally until it reaches a point B (T_1, P_B, V_B) taking Q_1 amount of heat. Work done, W_1 is equal to the area under the curve AB. Now the gas is isolated from the heat reservoir and is allowed to expand adiabatically and reversibly until it reaches a temperature T_2. This state is represented by a point C (T_2, P_C, V_C) on the T_2 curve. The work done W_2 will again be represented by the area under the curve BC.

We have, thus extracted some work from the system and have brought it to a state C (T_2, P_C, V_C) from a state A (T_1, P_A, V_A). Now let us try to restore the system to its original state so that it can be said to be working in cycles. From point 'C' if we return to 'A' via 'B' then we would be performing work on the system by an amount which is exactly equal to the work derived from the system in bringing if from A to C. In that case, the net work by the system will be zero and the system cannot work as heat engine. However, we can find another path to return to A via D by performing a lesser amount of work (as indicated by the area under the curve CDA). This can be done by bringing the gas at C into thermal contact with the heat reservoir at temperature T_2 and compressing the gas isothermally and reversibly until we reach the point D. Let the heat given out be Q_2 and work done on the system be W_3.

As we have seen earlier, if we compress the gas at D adiabatically and reversibly until temperature T_1 is reached we would reach the point A. Let the work done on the system be W_4.

Now the system has reached the original state and the area $ABCD$ represents the net work done by the system, which is not zero. Thus, the system is a heat engine, which has worked in a cycle in a reversible manner and has converted some heat into work. The cycle the system has undergone is called the "Carnot's Cycle".

5.2.1. Efficiency of Carnot's Cycle

Now let us examine the quantity of heat converted into work and try to compute the efficiency of the cycle in terms of temperatures T_1 and T_2.

If we examine the various quantities of work and heat at various stages, we come to the following conclusions.

Net work done in the full cycle,

$$W = W_1 + W_2 - W_3 - W_4$$

$$Q_1 = W_1 = RT_1 \ln(V_B/V_A)$$

$$W_2 = \int_{T_2}^{T_1} C_v dT = W_4$$

$$Q_2 = W_3 = RT_2 \ln(V_C/V_D)$$

Further, $P_A V_A/T_1 = P_B V_B/T_1 = P_C V_C/T_2 = P_D V_D/T_2$

also $T_1 V_A^{\gamma-1} = T_2 V_D^{\gamma-1}$

and $T_1 V_B^{\gamma-1} = T_2 V_C^{\gamma-1}$

Hence, $V_A/V_B = V_D/V_C$

or, $V_B/V_A = V_C/V_D$

and $W_3 = RT_2 \ln (V_B/V_A)$

Hence, total work done

$$W = W_1 - W_3 = RT_1 \ln (V_B/V_A) - RT_2 \ln (V_C/V_D)$$

$$= R(T_1 - T_2) \ln (V_B/V_A)$$

But $W = W_1 - W_3 = (Q_1 - Q_2)$

Efficiency of the engine

$$= W/Q_1$$

$$= (Q_1 - Q_2)/Q_1 = 1 - Q_2/Q_1$$

But since $Q_1 = W_1$, efficiency of the engine is also

$$= (R(T_1 - T_2) \ln (V_B/V_A))/(RT_1 \ln (V_B/V_A))$$

$$= (T_1 - T_2)/T_1 = 1 - T_2/T_1$$

Thus, the efficiency of reversible heat engine depends only on the temperatures between which the heat engine operates. The lower the temperature of sink where the lower quality of heat is discarded the higher is the efficiency of the heat engine.

From the above, we get,

$$1 - Q_2/Q_1 = 1 - T_2/T_1$$

or $Q_2/T_2 = Q_1/T_1$

or $Q_1/T_1 - Q_2/T_2 = 0$

5.3. ENTROPY

From this relationship evolves the concept of 'Entropy.'

The above expression (the relationship achieved over a cycle of reversible operation) is often stated in generalised form as

$$\oint (dQ_{rev}/T) = 0$$

The above implies that dQ_{rev}/T points to a new state function since its value would remain independent of the path the system would have taken to reach that state. This new function, normally designated S, is exactly defined only mathematically and in differential form.

$$dS = dQ_{rev}/T$$

The concept of Entropy would be discussed in detail in the next chapter.

Chapter- 6

Entropy, Third Law and Feasibility

6.1. ENTROPY

Entropy is a Greek word meaning 'change.' Change has been interpreted as increased disorder since increase in Entropy has been seen to be associated with increase in disorder or randomness.

Entropy 'S' has been defined by the following mathematical relationship

$$dS = dQ_{rev}/T$$

Since,

$$\oint (dQ_{rev}/T) = 0$$

therefore, after performing a cycle of operation reversibly, and on reaching back to the initial state, the change in entropy is zero. That is, the entropy of system does not change over a complete reversible cycle.

Therefore, S is a state property or an exact differential. Entropy cannot be easily defined but can be described in terms of entropy increase accompanying a particular process.

In an infinitesimal stage of an appreciable process, the entropy increase, dS is given by heat taken up isothermally and reversibly divided by absolute temperature T at which it is absorbed.

6.2. SIGNIFICANCE OF ENTROPY

In the discussion in the last section on Carnot's Cycle, we have noted that for a full cycle of operation of a reversible heat engine

$$Q_1/T_1 = Q_2/T_2$$

The terms Q_1 and Q_2 are, by definition, the quantity of heat transferred into the system isothermally at T_1 temperature and heat transferred out of the system at T_2 temperature, respectively, both operations being performed in reversible manner.

Thus, Q_1/T_1 is the integration of $dQ(\text{rev})/T$ during the first part of the cycle and so is Q_2/T_2 for the third part of the cycle. So these values are, in fact, entropy changes of the system during the first part and the third part of the cycle. The second and fourth parts being adiabatic and reversible, there are no entropy changes associated with these parts. The entropy change in the first and the third part being equal and of opposite sign, the entropy change in the full reversible cycle is zero. So, even though the reversible heat engine takes up heat from heat source, performs work and discards unutilised heat, on completing the cycle of operation there is no change in entropy.

The working of any cyclic reversible heat engine, through whatever path it operates, can be broken down into a large number of Carnot's Cycle (Fig. 6.1). Therefore, as long as the process is carried out reversibly, any heat engine will be converting heat into work with the same outcome and without any net change in entropy.

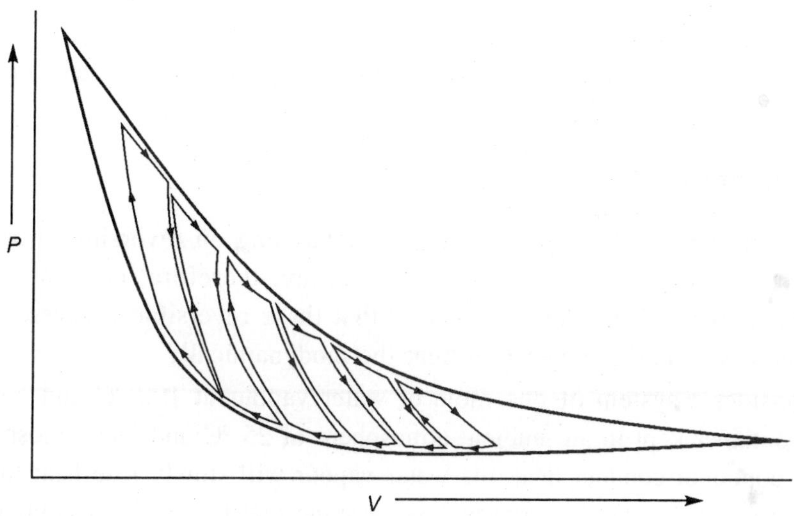

Figure 6.1 Any cyclic reversible heat engine can be seen as comprising a large number of Carnot's Cycles

6.3. THERMODYNAMIC TEMPERATURE SCALE

If a reversible heat engine operates between two constant temperature reservoirs, then the thermodynamic temperature is defined as being proportional to the quantity of heat transferred to and from it in a reversible cycle (*Kelvin, 1848*).

6.3.1. Zero of the Thermodynamic Scale

It is the lower temperature of a reversible heat engine cycle with an efficiency of unity (*i.e.,* one capable of converting heat completely into work).

It has been a dream of many scientists to reach as close to this temperature (0°Kelvin or 0° Rankine, approximately equal to –273.2 °C) as possible (kindly refer to Chapter 1).

6.4. NON-CYCLIC PROCESS

What happens when the conversion of heat into work is not cyclic. Let us take only part one of operation, represented by curve AB in Fig. 5.2. Here Q_1 heat has been taken up by the ideal gas in piston (Fig. 4.1) from the heat source at temperature T_1. Corresponding work done is $RT_1\ln(V_B/V_A)$. Entropy of the system (gas in piston) has increased by Q_1/T_1

or, $$\Delta S_G = Q_1/T_1$$

At the same time, the heat source being at the same temperature and having lost an equal amount of heat reversibly has also lost an equal amount of entropy.

$$\Delta S_H = -Q_1/T_1$$

In the system of the gas in piston and the heat source, taken together, there has been no change of entropy. There has not been any other entropy change any where else in the surroundings. Thus, the entropy of the system and all the surroundings taken together has not changed. Thus, we see that, in any reversible process, entropy of a system may change but that of universe does not change.

6.5. REAL PROCESS

Readers might be wondering why we are emphasising on reversible processes, which, by nature, must take infinite time to complete and are, therefore, not real processes. But readers must have guessed the answer as well; that these reversible processes are idealised forms of real processes and are easier to treat thermodynamically.

Let us consider a system of one mole of water vapour at 100 °C and one atmosphere pressure in a container kept in an ambient atmosphere of 25 °C and one atmosphere pressure. If the walls of container conduct heat, the water vapour will slowly lose heat to surroundings and condense to water. Let us, for convenience, assume that the walls of container expand or contract with change of pressure. In real situation, there would be no such wall.

This is an example of a real process and it is not a reversible process since infinitesimal change in constraints cannot bring back the previous state of the system. How to calculate entropy change in such a process?

Let us consider that the same change is brought about in a reversible manner employing a reversible heat engine to attain the same end. Let the ideal gas contained in a cylinder and a piston in Fig. 4.1 act as the heat engine, which would extract heat from the system of one mole of steam at 100 °C and 1 atmosphere pressure. This ideal gas will have to be at 100 °C, it would pick up the latent heat of condensation of steam and expand doing work on the surroundings. Thereafter, we allow the ideal gas system to expand adiabatically and reversibly until its temperature reaches 25 °C and thereafter it can discard heat reversibly through isothermal compression, the work now being done by surroundings on the ideal gas. After transferring out equal amount of heat as it had gained from the system of steam, the ideal gas system is then compressed adiabatically and reversibly until it reaches 100 °C. But the ideal gas system has not reached the same state as it had started with, since its entropy has changed (point E in Fig. 6.2).

Let Q_1 be the heat extracted from the system of steam

The entropy lost by the system of steam

$$= Q_1/373$$

Entropy gained by the system of ideal gas during isothermal expansion stage is also

$$= Q_1/373$$

But entropy lost by this system during isothermal compression stage

$$= Q_1/298$$

The same amount of entropy is gained by the surroundings.

Thus, entropy lost by the system of ideal gas

$$= Q_1/298 - Q_1/373$$
$$= Q_1(1/298 - 1/373)$$
$$= Q_1*(373 - 298)/(298*373)$$
$$= Q_1*75/(298*373)$$

And the surroundings and the system of steam together gain the same amount of entropy. The entropy of universe thus remains constant.

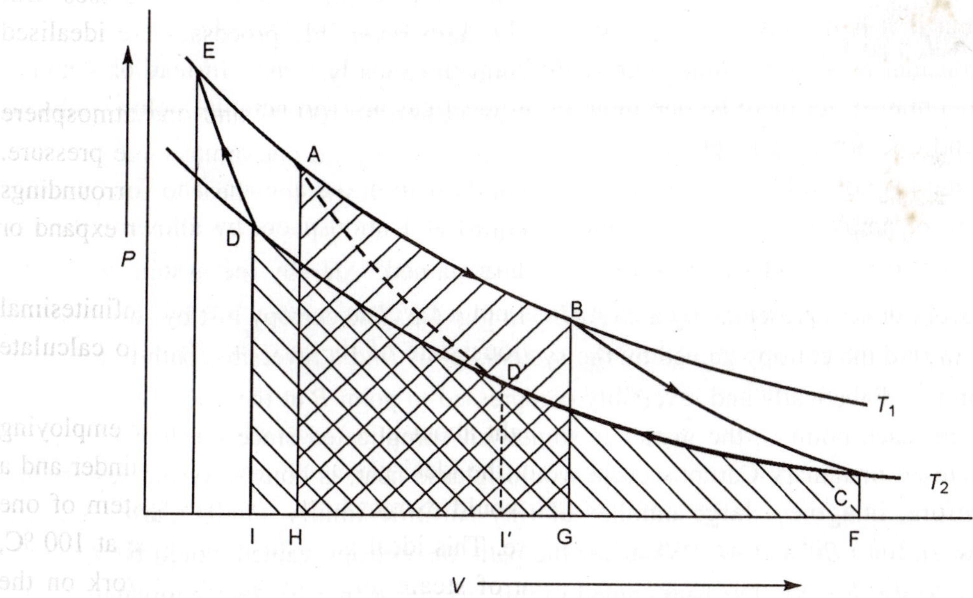

Figure 6.2 An ideal gas system assisting in simulating a real process in a reversible manner

6.6. ENTROPY CHANGE IN REAL PROCESS

Coming back to the example of real process where the steam has condensed and the latent heat has been dissipated into the surroundings. No ideal gas system exists in between and the change has taken place – not reversibly – but spontaneously or irreversibly. But since entropy is a state property, it does not matter what path has been taken to reach a particular state. The states of steam and the surroundings are exactly same as that in the example of reversible change using ideal gas, after the heat transfer has taken place. Only the system of ideal gas is absent.

Thus, in the example of real process, the system of steam has lost an entropy of $Q_1/373$ while the surroundings have gained an entropy of $Q_1/298$. Thus, there has been an overall gain of entropy of $Q_1*75/(298*373)$ by the universe.

By similar examples, it can be proved that in an irreversible process (or a real process), the system under consideration may lose or gain entropy, but the entropy of universe always increases!!!

6.7. PROCESS FEASIBILITY

Reasoning the other way around, if for a proposed process we are able to measure the entropy changes of both system and surroundings and we find that the sum total is positive, we conclude that the process is feasible. This appears to be a tall order, *i.e.*, to compute entropies of both system and surroundings. But a way out has been found as we will find in the latter sections.

If the entropy change of universe is found to be zero we conclude that the system is in equilibrium, but if it is negative we conclude that the process is not feasible.

Let us consider another example slightly different from the last one. Instead of steam at 100 °C in the container, let there be one mole of an ideal gas at a temperature below ambient, say, at 0 °C and one atmosphere pressure.

If the system is not isolated it would slowly gain heat from the ambient and finally attain the temperature of ambient, the pressure being retained at 1 atmosphere by allowing the gas to expand against the atmospheric pressure. The heat gained, ΔH, by the system would be equivalent to work done represented by area *AEFC* in Fig. 6.3. The entropy lost by surroundings is $\Delta H/T$. But to find the entropy gained by the system, let us find a reversible path to the point *C*. The gas can be adiabatically and reversibly compressed to point *B* at the line of temperature *T*. Thereafter, to reach point *C*, the matter is somewhat complicated since the temperature of heat sink, where an imaginary Carnot's cycle would release heat, is constantly increasing. We have to, therefore, imagine a large number of tiny (infinitesimally small) Carnot's cycle operating between lines *BC* and *AC*. Whatever the path, the entropy gained would be $\Delta S = S_C - S_B = S_C - S_A$. Since $S_A = S_B$. The heat gained in this manner would be $T\Delta S$ equivalent to the work done represented by area *DBCF*.

There are other ways of determining values of entropy at different states such as at *A* and *C* (S_A and S_C) and for commonly used materials and conditions, S_A and S_C may be available from Tables. Thus, ΔS is separately determinable.

In this example, entropy gained by the universe would be

$$\Delta S_{\text{univ}} = \Delta S - \Delta H/T$$

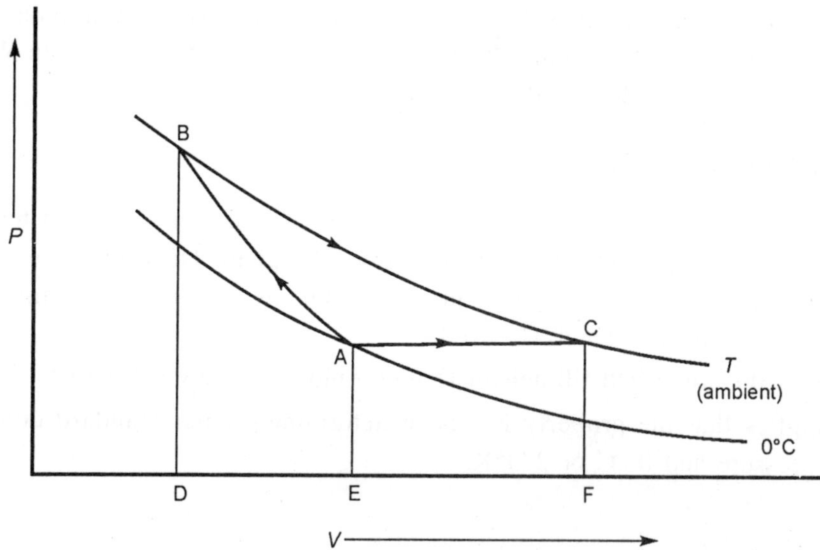

Figure 6.3 Another example of a real process: entropy gained by the system is more than lost by the surroundings

6.8. FREE ENERGY

We see now that, in cases of the type cited above, the entropy change of universe is determinable based on the properties of system itself. We need not really study the surroundings to find out its entropy change. In fact, we can simplify the above expression to

$$-T\Delta S_{\text{univ}} = \Delta H - T\Delta S$$

Or we define a term ΔF, change in Free Energy (of the system) so that

$$\Delta F = -T\Delta S_{\text{univ}} = \Delta H - T\Delta S$$

ΔF in this example has a negative value, and it has dimension of energy, similar to heat energy. It signifies an energy which can be easily released, other than the $T\Delta S$ which signifies bound energy, not easily accessible. For a real process, ΔF is negative. When ΔF is zero, then change in the entropy of universe is zero and we are looking at a process which is operating reversibly and hence is always at equilibrium. If ΔF is positive, such a process is not feasible.

We have described above a constant pressure process, and we use the alternate term Gibbs Free Energy since, for constant pressure processes, Gibbs Free Energy is applicable. Gibbs Free Energy, F is defined mathematically as

$$F = H - TS$$

For constant volume process, the similar property, which is a measure of feasibility of processes, is the Work Function or the Helmholtz Free Energy, A, which is mathematically defined as

$$A = E - TS$$

For constant volume processes, such as catalytic synthesis of ammonia from nitrogen and hydrogen in pressurised (about 300 atmospheres) reactor, the change in Work Function would be a measure of the feasibility of reaction.

6.9 ELLINGHAM DIAGRAM

Originally, a plot of free energy of formation of oxides, plotted against temperature was referred to as the Ellingham's Diagram, but now Ellingham Diagram have been made for sulphides and many other groups of compounds. But, by far the Ellingham's Diagram for oxides has been the most extensively used.

Fig. 6.4 is the simplified Ellingham Diagram plot for a few key oxides. The superscript '°'on ΔF implies that the property has been determined under standard conditions – one atmosphere pressure and 0 °C or 273°K.

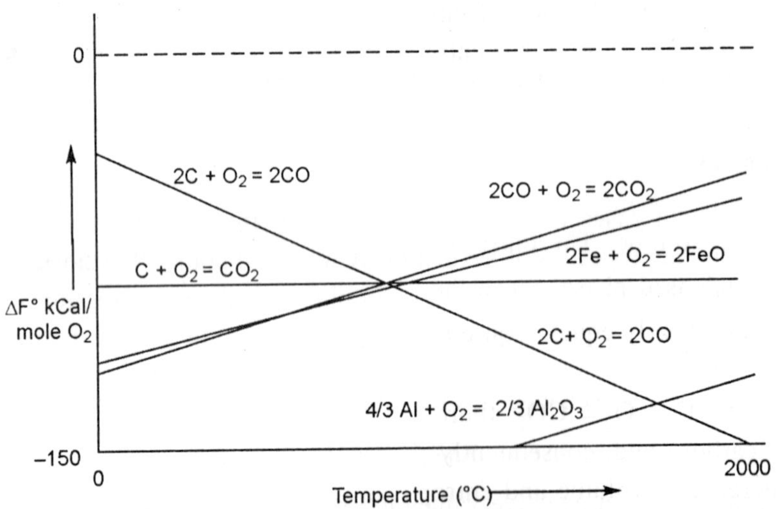

Figure 6.4 Elllingham diagram for a few oxides

These plots are near straight lines since $\Delta H°$ and $\Delta S°$ are nearly constants ($\Delta F° = \Delta H° - T\Delta S°$). There are points of slope changes at the temperatures of transformation (of metal and metal oxide) – these are not shown in the sample plot. The slope change is marginal when there is solid to liquid transformation, but is pronounced when transformation to gas occurs.

Only two metal oxide lines ($2Fe + O_2 = 2FeO$ and $4/3\ Al + O_2 = 2/3\ Al_2O_3$) are shown in the sample diagram (Fig. 5.4). In reality there are a large number of nearly parallel lines (for, say, CaO, MgO, SiO_2, BaO, K_2O, Na_2O, etc.). All these lines are plotted on the basis of involvement of one mole of oxygen utilisation. e.g.

$$2Ca + O_2 = 2CaO$$

$$2Mg + O_2 = 2MgO$$

$$Si + O_2 = SiO_2$$

$$4/3\ Al + O_2 = 2/3\ Al_2O_3 \text{ etc.}$$

This is done to ensure ease of algebraic addition and subtraction of the chemical equations. Stated differently, we can conclude that the lines lying lower in the diagram denote a more stable oxide. An element, *e.g.*, Al, capable of giving a more stable oxide can reduce a relatively less stable oxide (*e.g.*, FeO). The reaction between Al and FeO is the basis of the Thermit Welding Process.

By a similar logic it can be seen that carbon is poor reducer of metal oxides at lower temperatures but can reduce most of the metal oxides at higher temperatures, since *CO* line has a negative slope and cuts across most of the other metal oxide lines.

Volume change is an indication of entropy change since higher volume signifies a state of higher disorder and hence of higher entropy. And if the volume change is large, this would almost entirely account for the change in entropy. Volume changes in solid to solid or solid to liquid transformation becomes almost of zero significance when a gaseous constituent either appears or is consumed. In cases where metal and its oxide both are in condensed phase (solid or liquid) form, one mole of gaseous oxygen is consumed and converted to a condensed phase of negligible volume. The entropy therefore decreases – almost equally for all the metal oxides in condensed phase. Since the slope of this plot is $-\Delta S°$, the slope of all such lines are positive and nearly equal. Consequently, the metal oxide lines are nearly parallel and rising.

In the equation $C + O_2 = CO_2$, there is no change in volume (we are neglecting the volume of solid carbon) and, consequently the line is horizontal. In the equation $2C + O_2 = 2CO$, there is increase in volume and entropy, and therefore, the slope is negative. This line cuts through most of the metal oxide lines and therefore carbon has proven itself as an effective reductant for most of the metal oxides.

Rising lines indicate that oxides of the more noble metals like Ag_2O, Au_2O, etc. would decompose, on its own, to metal simply by heating. The diagram also indicates that C can reduce even hard-to-reduce oxide like Al_2O_3, at about 1900 °C. This fact is of little commercial significance at the moment, since cost-effective means to curb the reversal of reaction during cooling has not yet been found.

However, readers may note that the diagram has been drawn at standard state (one atmosphere pressure). Inferences drawn from the diagram may change depending upon how far it is possible to stretch the system away from standard state. For instance, as per this diagram, it does not appear possible for CO to reduce FeO to Fe (the '2Fe + O_2 = 2FeO' line lies above '2CO + O_2 = $2CO_2$' line). But it is being done regularly when sponge iron is made (either in rotary kiln or in shaft) by maintaining a large excess of CO over CO_2.

6.10. THIRD LAW

Studies related to entropy changes revealed that reduction in temperature leads to decrease in entropy change for all processes. It was therefore, postulated that for a process occurring at absolute zero temperature the entropy change would be zero. This has led to a basis from which absolute values of entropy can be determined, taking entropy at absolute zero of temperature to be zero. Thus, unlike *E*, *H* and *F* whose changes can be accurately measured but not the absolute value, the absolute value of entropy can indeed be measured. We take, for calculation purposes, enthalpy of elements in a defined standard state to be zero, but that assumption is only for convenience, no molecule or atom can have zero heat content at ambient conditions. On the other hand, a fully ordered (crystalline) solid at absolute zero temperature will have zero entropy.

6.10.1. Absolute Entropy and Standard State

After the formulation of Third Law, it is possible to determine the absolute values of entropy of different solid crystalline elements with the aid of knowledge of its C_p data over the entire temperature range

$$S_T^\circ = \int_0^T (C_p/T)\,dT$$

The value of S_T° (entropy at temperature *T* and at standard pressure *i.e.*, 1 atmosphere pressure) is best determined graphically by plotting C_p/T against *T* and measuring the area underneath the curve (Fig. 6.5).

Based on knowledge of heats and temperatures of transformations, entropies of liquid, gaseous and other transformed phases can be determined. Similarly, it is possible with the knowledge of heats of formation and temperatures of equilibrium to determine entropy in many cases. In case of stable compounds, the integration method (Fig. 6.5) is used for entropy determination.

While absolute entropy values can now be determined absolute values of Internal Energy and Enthalpy cannot be conceived. For ease of calculation, related especially to metallurgical reactions (constant pressure processes), a suitable reference point of enthalpy is conventionally chosen and that is – for pure elements, the enthalpy is zero when in 'Standard State'. Standard

State is defined as 25 °C or 298 °K and atmospheric or 1 bar – 760mm of Hg – pressure, *i.e.*, close to ambient conditions.

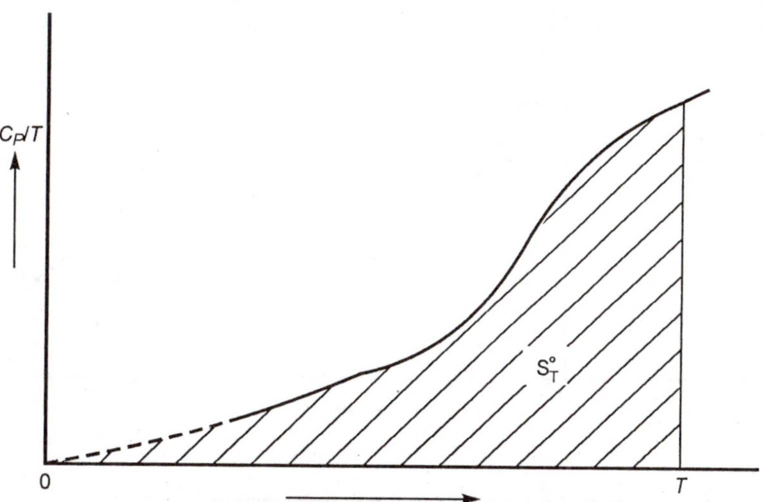

Figure 6.5 Graphical computation of entropy

6.11. EXAMPLES OF ALTERNATE HEAT ENGINES

We have derived the validity of the free energy concept using examples of mostly physical changes only, but it is equally applicable in cases of chemical change. We have so far avoided proof of applicability in chemical changes simply because it is somewhat difficult to conceive of a thermodynamically reversible route for most of the common chemical reactions.

As an example we would try to conceive of a Carnot's cycle without using ideal gas as the system. Let the container in Fig. 4.1 contain water and steam in equilibrium instead of an ideal gas. Here pressure would be 1 atmosphere if temperature was 100 °C. If heat is supplied to the system, more of water would get converted to steam. It is easy to conceive that, if heat is transferred infinitesimally slowly, heat transfer can be carried out in thermodynamically reversible manner with corresponding increase in volume. By releasing pressure, also in a reversible manner, more of water gets vaporised and volume increases further to point C (Fig. 6.6) at reduced pressure and temperature. Thereafter the system can be made to lose heat reversibly at the lower temperature, which would make some steam to condense to liquid water with reduction in volume.

After reaching point D, the system can be adiabatically compressed to point A with further condensation of vapour into liquid water.

The system has thus worked in a cycle and has performed work equivalent to area ABCD; it had taken up heat equivalent to area ABFH at higher temperature T_1 and discarded heat equivalent to area CDGE at lower temperature T_2.

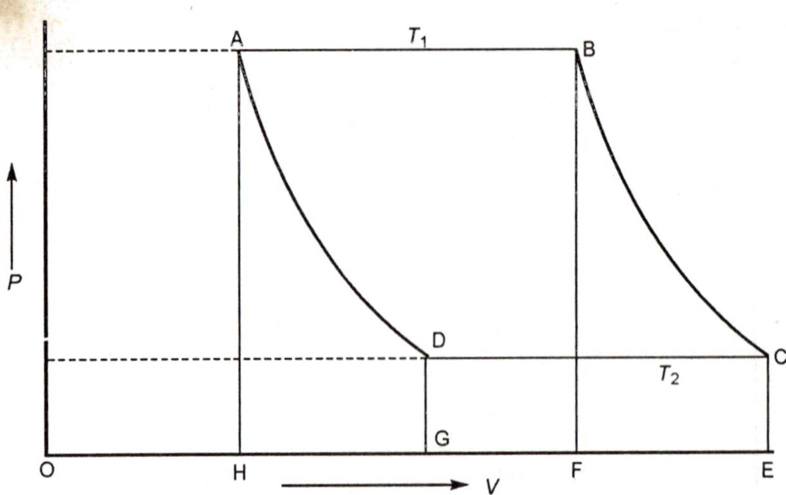

Figure 6.6 Carnot's Cycle with water vapour system

Let us take another example. A system of CO and O_2 gases would tend to combine to form CO_2, but tendency for a reverse reaction tends to increase at higher temperatures. At certain temperature (in excess of 2000 °C) and pressure conditions all the three constituents would be coexisting ($2CO + O_2 \leftrightarrow 2CO_2$). The forward reaction is accompanied with liberation of heat and decrease in volume and vice versa.

Under a condition when all the three constituents exist in significant quantities and are at equilibrium, if a certain quantity of heat is supplied at constant pressure, more CO_2 would tend to break up and form CO and O_2 causing increase in volume. The system would be doing work on surroundings. Thereafter, if the system is allowed to expand adiabatically and reversibly, its temperature would drop causing formation of more CO_2 from CO and O_2. Both volume increase and temperature drop would tend to be compensated by the chemical combination of CO and O_2, but none the less there would be a net increase in volume and decrease in temperature. Taking out heat isothermally from system while compressing it reversibly – and thereafter compressing it adiabatically and reversibly – would give a plot similar to that illustrated in Fig. 6.6 although with different slopes. This system involving chemical change can also work as a heat engine just like the previous example of physical change. The major difference in the two examples is that in the example of physical change, the process is endothermic in the forward direction, while in the chemical reaction example, the forward process is exothermic.

Thus, we can see that heat engines can be conceived with not only system of an ideal gas but also with systems involving physical and chemical changes. And Carnot's Theorem (2nd Law) holds for all such examples.

Chapter- 7

Solution Thermodynamics

Equilibrium condition among the pure substances is represented by a situation where thermodynamic reversibility is maintained. On the other hand, the feasibility of reaction is represented by the extent of deviation from thermodynamic reversibility – the Free Energy change is a measure of this deviation. In real life situation, the extent of deviation from thermodynamic reversibility assumes greater importance. Therefore, 'Free Energy changes in real systems' is a very important subject of study. While Free Energy changes in cases where both reactants and products are in pure state, are easy and convenient to study, systems become somewhat complex when one or more of these constituents are held within a solution. For systems involving solutions, new concepts of Fugacity, Activity and Partial Molar properties have been evolved.

7.1. CHEMICAL POTENTIAL, FUGACITY AND ACTIVITY

In a system involving physical change, the system has at least two phases – each of these can be viewed as a subsystem. Since the two subsystems are in equilibrium with each other, the free energies of the two phases must be equal.

$$\Delta F_v = 0 = \Delta H_v - T_v \Delta S_v$$

or

$$\Delta S_v = \Delta H_v / T_v$$

Thus, there is difference in entropies of the two phases depicted by the term ΔS_v above even though the two phases are in equilibrium; *i.e.*, the phases have the same quantitative tendency to interact with a third phase. This tendency to interact or react has been referred to as 'potential' (*i.e.*, ability or dormant ability to react). In quantitative terms, this potential is nothing but the free energy of the phase.

With reference to the tendency to react chemically, this 'potential' is referred to as 'Chemical Potential.' In such situations, the term 'Chemical Potential' is used synonymously with free energy.

7.1.1. Solutions and Activities

A common example of a solution is a homogeneous gas mixture, and presence of such homogeneous gas mixture often alters the feasibility situation. From Fig. 6.4 (Ellingham Diagram), we can see that FeO formation line lies slightly below the CO_2 formation line implying that an equal mixture of CO and CO_2 would not be able to reduce FeO. But by maintaining a large excess of CO over CO_2, the sponge iron making process has become a massive industrial success.

7.1.2. Activities in Liquid Solution

If we mix water and alcohol (C_2H_5OH, ethanol), they dissolve to form a single liquid phase. The partial pressures of the individual vapours get reduced in this process compared to what they were, when the constituents were in pure state. The tendencies to react of both these constituents also goes down compared to the level of reactivity when they were in pure state. We say that the 'activities' of the constituents decrease when they go into solution.

We would note that there is some relationship between the vapour pressure and the tendency to react (reactivity or, since chemical process is not always the consideration, we just say 'activity'). In case of an ideal gas, the pressure is synonymous with activity. The more we compress CO_2 gas the more it goes in solution in water (or in aerated drinks). Similar is the case with gases like oxygen and nitrogen, either in contact with water or with a molten metal such as liquid iron.

The above fact gives us a quantitative basis for measurement of thermodynamic activity of a constituent. As we would like to use this concept of activity in relation to a reference (standard) state, we define activity at the standard state to be unity. Thus pure ideal gases at 25 °C (298 °K) and at 1 atmosphere pressure are assumed to have unit activity. At this temperature, if they were kept at half atmospheric pressure, they would have an activity of 0.5.

For a pure substance in condensed phase (solid or liquid) their vapour pressure is taken to be related to their activities. When we measure equilibrium vapour pressure, the vapour and the solid (or liquid) are at equilibrium and hence the 'activities' of the constituent in both phases are the same. Hence, the activity of the vapour phase would represent the activity of the solid or liquid.

We know that even non-ideal gases behave like ideal gases when reduced to low pressures. As the vapour pressures of majority of solids and liquids of interest in metallurgical processes are very low, these vapours can be assumed to behave like ideal gases.

The vaporisation tendency of condensed phases has been thermodynamically referred to as "Fugacity" or escape tendency. After all, the molecules, which join the vapour phase, are trying to 'escape' away from the condensed phase. Fugacity is more correctly defined mathematically and this is presented in a latter section.

The 'activity' of a condensed phase is defined as its 'fugacity' relative to the fugacity at its standard state, *i.e.,* pure constituent at 25 °C (298 °K) and 1 atmosphere pressure.

7.1.3. Equilibrium in a Single Component System

Let us consider a pure substance say, water, at ambient conditions. Water naturally evaporates, and its equilibrium vapour pressure is equal to the pressure of vapour when water is kept in an evacuated chamber. Since equilibrium condition exists, free energy of transformation is zero or, stated differently, the free energies of the two phases are same. Since it is the case of a single pure substance, we say that the "activities" of the two phases are same.

7.1.4. Two-Component System and Partial Molar Properties

When two or more components mix to form a solution, naturally the properties of the individual constituents get affected. In the example of water-alcohol mixture, if absolute (100%) alcohol is diluted with water, the vapour pressure of alcohol decreases almost proportionally to the concentration of alcohol in the solution. On the other hand, the equilibrium vapour pressure of water starts increasing almost (but not exactly) proportionally to the concentration of water in the solution. The pressure exerted by the vapours of the solution is therefore made up of the partial pressures of alcohol and water. Thus, any property of the solution is made up of "partial properties" of the constituents in the solution, which we denote with a bar (–) on top.

$$P = \overline{P}_1 + \overline{P}_2 + ...$$

In case of an extensive property, like volume, the same relationship holds. In such cases, we are generally interested in the per mole contributions of property of the individual constituent. Thus,

Gross volume = Sum of volume contribution of individual constituents

or $$(n_1 + n_2 + ...)\, V = n_1\overline{V}_1 + n_2\overline{V}_2 + ...$$

or $$V = N_1\overline{V}_1 + N_2\overline{V}_2 + ...$$

where,

n_1, n_2, n_3, ... etc. are the number of moles of each constituent

and

N_1, N_2, N_3, ... etc. are the mole fractions of each constituent

These values of $\overline{V}_1$, $\overline{V}_2$, $\overline{V}_3$, etc. are not normally same as the values of V_1, V_2, V_3, etc., *i.e.,* the volumes of the individual constituents originally used to make up the solution. We thus evolve the concept of partial molar property (in this case partial molar volume) $\overline{V}_1$, $\overline{V}_2$, $\overline{V}_3$, etc.

7.1.5. Ideal Mixing and Positive and Negative Deviations

Graphical computation of partial molar volumes for a two-component system has been illustrated in Fig. 7.1. Here, the volume of one mole of the mixture (specific volume) has been plotted against the mole fraction of constituent B, in a mixture of constituents A and B. By drawing tangent at any point on the curve, one can obtain from the intercepts the molar (*i.e.*, per mole) contribution of the individual constituents i.e. V_A and V_B.

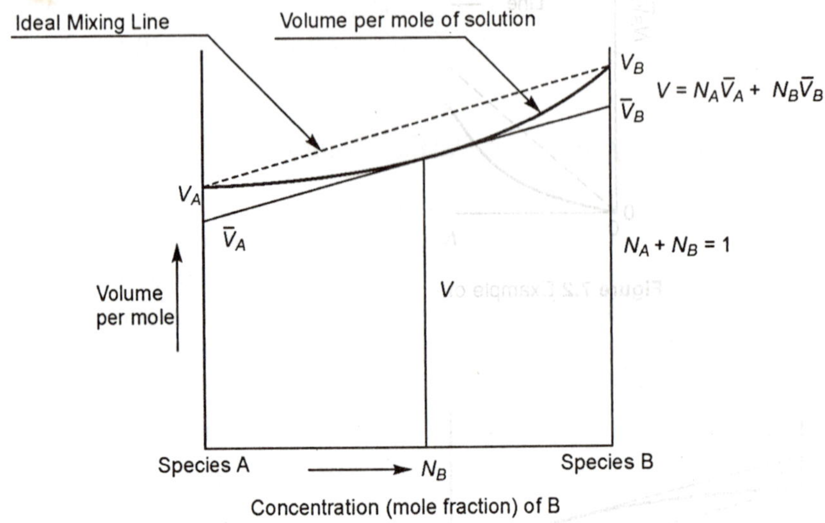

Figure 7.1 Graphical computation of partial molar volume in a 2-component system

If $\bar{V}_A$ would, at all points, have been equal to V_A (*i.e.*, the curve would merge with the straight line between V_A and V_B and, therefore, $\bar{V}_B$ at all points would be equal to V_B), we say that the system is ideal and this is a case of ideal mixing. Examples of ideal mixing are more common in solid solutions, such as in Ag-Au and Au-Pt systems. The present example (Fig. 7.1) is a case of 'negative deviation' from ideality. The partial molar property values are lower than those for pure components. In Fig. 7.2, the volume contribution of component $B(= N_B \bar{V}_B)$ has been plotted against mole fraction B. The line keeping below the diagonal in this Figure is a typical representation of a system exhibiting negative deviation from ideality. Here the molecules of A and B have stronger affinity for each other as compared to the affinities between molecules of A-A and B-B. In fact the A-B affinity is more than the linearly interpolated (averaged) value of affinities between A-A and B-B.

Similar to the above, Fig. 7.3 represents a case of positive deviation from ideality. In the first part of the figure, the specific volume *i.e.*, volume occupied by one mole of the mixture has been plotted against the mole fraction of the second species. In the second part of the figure, the volume contribution of the second species has been plotted against mole fraction.

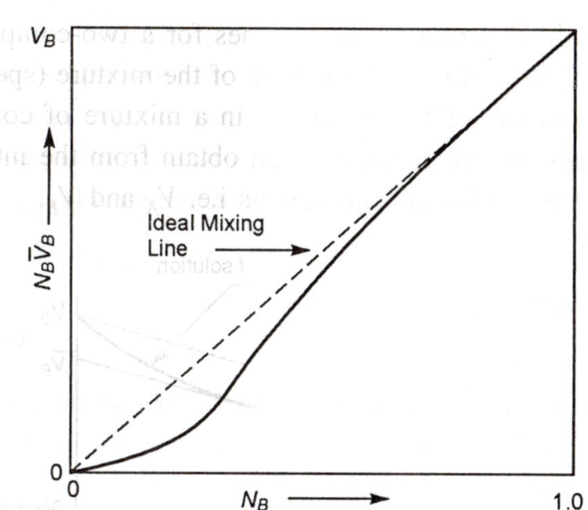

Figure 7.2 Example of negative deviation from ideality

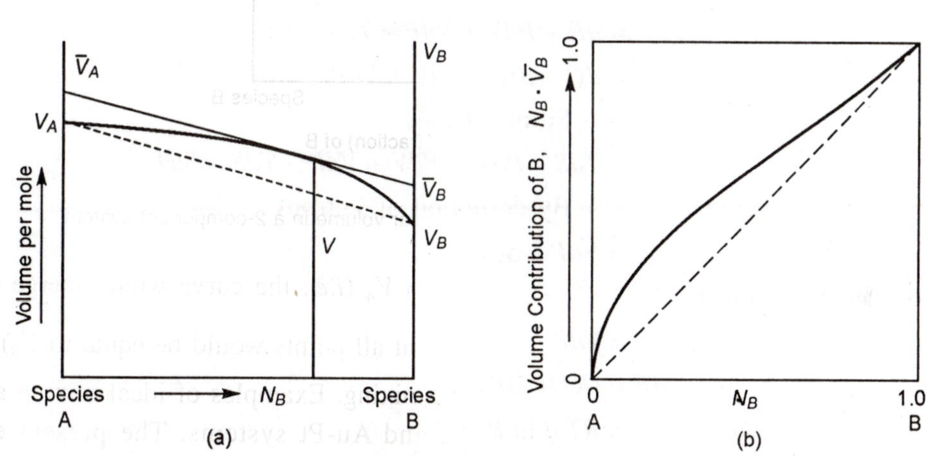

Figure 7.3. Example of positive deviation from ideality

It can be noted from the examples of both negative and positive deviation of ideality for Figs. 7.2 and 7.3(b) that when mole fraction approaches the value of one, the volume contribution line tends to merge with the ideal mixing line. We say that at higher concentrations the species follows Raoult's Law. On the other hand at very low concentrations, the slope of the line is very much different from the ideal mixing line. But to start with line is nearly linear. Here, although the volume contribution is not equal to the mole fraction of the species, it is proportional, at least in a small and very dilute range. We say that the species follows Henry's Law.

7.2. RAOULT'S AND HENRY'S LAWS

Readers would have guessed that in case of ideal mixing, the system is said to follow Raoult's Law. And in the regions where the volume contribution is proportional to its mole fraction the system is said to follow Henry's Law. This is true for all other thermodynamic property of relevance to our study. In metallic systems as well as in metal oxide systems (such as slags – metallurgical and refining slags), properties of greater relevance are vapour pressures, fugacities and activities.

7.2.1. Fugacity – Idealised Partial Pressure

For any pure component in a condensed phase, whether solid or liquid, kept at steady conditions for sufficient time, can be considered to be in equilibrium with its vapour. For metallurgical systems, where vapour pressures are extremely small, close to equilibrium conditions are easily achieved in a very short time. The free energy change of conversion of condensed phase to vapour can be denoted as follows:

$$dF = d(H - TS) = d(E + PV - TS)$$

– By definition of Free Energy

$$= dE + PdV + VdP - TdS - SdT$$

$$= dQ - dW + PdV + VdP - TdS - SdT$$

– From 1st Law

$$= TdS - PdV + PdV + VdP - TdS - SdT$$

– By definition of Entropy

$$= VdP - SdT$$

At constant temperature

$$dF = VdP$$

or

$$dF = (RT/P)dP$$

$$= RT\, d \ln P$$

Here P naturally refers to the vapour pressure of the component.

No wonder the vapour pressure assumes a high importance in thermodynamic study of metallurgical systems. In the above derivation, the applicability of ideal gas law has been assumed. It would be very nearly true for vapours of metals and their oxides, as their pressures are extremely low. For a more general applicability, the above relationships have been restated in terms of 'fugacity' or escape tendency, which has been qualitatively introduced in the beginning of this chapter. Quantitatively, 'fugacity, f' is defined as

$$dF = RT\, d \ln f$$

Naturally, 'f' becomes equal to 'P' *i.e.*, partial pressure when ideal gas law is applicable. In other words, fugacity is the idealised partial pressure.

7.2.2. Activity – Relative Fugacity

Similarly, the term 'activity, a' is defined as the ratio of its fugacity, f, to its fugacity at standard state, $f°$. Standard state is generally taken to be pure substance at standard temperature (298 °K) and pressure (one atmosphere pressure).

$$a = f / f°$$

Naturally, by this definition, activity at standard state of any constituent is unity.

If we integrate the earlier expression within the limits of standard state and the given state we get,

$$\int_{F°}^{F} dF = RT \int_{f°}^{f} d \ln f$$

or,

$$F - F° = RT (\ln f - \ln f°)$$
$$= RT \ln (f/f°)$$
$$= RT \ln (a)$$

7.2.3. Activity Coefficient and Henrian Activity

We have earlier given example of how the volume contribution of a constituent is related to its mole fraction, sometimes depicting positive deviation and sometimes negative deviation from ideality; while there are some instances where the system exhibits ideality thus following Raoult's Law with respect to that property.

If we consider thermodynamic activities in binary solutions we once again get similar patterns. In cases where Raoult's Law is obeyed, with respect to activity of the particular constituent, then activities are equal to mole fraction ($a_B = N_B$). Even if Raoult's Law is not obeyed, at very high mole fraction level (N_B approaching unity), the actual activity line merges with the ideal mixing or Raoult's Law line (Figs. 7.2 and 7.3(b)).

At very dilute concentration range and in cases where Raoult's Law is not obeyed, activity is found to be proportional to the mole fraction. As stated in Sec. 7.2, the system obeys Henry's Law. The proportionality constant is referred to as the Activity Coefficient (γ_B), of the constituent in the particular system.

$$a_B \propto N_B$$

or

$$a_B = \gamma_B N_B$$

thus,

$$\gamma_B = a_B / N_B$$

In fact, it is a general expression for any composition, *i.e.*, for any value of N_B. Only γ_B varies over the entire composition range. At very small values of N_B the constituent normally obeys Henry's Law and γ_B acquires a constant value.

In reactions involving refining of liquid metals in general, and during steel making process in particular, we deal with removal or addition of small quantities of constituents from or to an iron rich melt, the constituents in question are expected to obey Henry's Law. It was found to be convenient to define a new standard state – a 1-weight % solution of the constituent (solute) in the solvent (liquid metal, liquid iron in the above example). Often, this new activity term is referred to as the "Henrian Activity."

7.2.4. Interaction Coefficient

Using such a standard state, the thermodynamic feasibility calculations were expected to become much easier as only the value of the weight percent of the constituent was required to be used in place of activity values. But, at the same time, values of thermodynamic properties would need to be obtained as per this new standard and tabulated.

The above case was OK for approximate calculations, but not for precise calculations. It was found that activity-composition relationship was rarely linear up to 1 weight % of the constituent.

Let us define a term $\gamma°$, an activity coefficient, which is defined thus,

$$\gamma° = \text{Lim}\,(a/N) \quad \text{as} \quad N \to 0$$

In other words, $\gamma°$ is the activity coefficient at infinite dilution.

If Henry's Law is obeyed up to 1 weight percent of the solute constituent, γ at 1 weight percent solution would be same as $\gamma°$, otherwise $\gamma/\gamma°$ will have a value different from unity.

Wagner (*1952*) found that the following relationship holds reasonably well for almost all dilute solutions of metallurgical interest

$$\log(\gamma/\gamma°) \propto N$$

In other words

$$\log(\gamma/\gamma°) = \in N$$

$\in$ is referred to as the interaction coefficient. In fact it is the self interaction coefficient of that constituent. Needless to say that if Henry's Law is obeyed up to 1 weight percent concentration, $\in$ would have a zero value. Otherwise, it would have a definite (positive or negative value).

In the above relationship, the left-hand side term, being ratio of two activity coefficient terms is independent of the standard state chosen. The activity coefficient (generally termed f) based on 1 weight percent solution as the standard state, can very well be used. On the right hand side, N or the mole- (or atom-) fraction, would be replaced by weight per cent of solute and the interaction coefficient would have to absorb within it the corresponding conversion factor for composition. In this form, the interaction coefficient is generally represented by the symbol 'e'.

Up to this point, no definite advantage can be perceived by the use of a 1 weight percent solution as the standard state. But if more than one solute is present, Wagner (*1952*) found that the activity coefficient is changed by the following expression

$$\log f_A = e_A{}^A\,(\%A) + e_A{}^B\,(\%B)$$

where

$e_A{}^A$ = Self interaction coefficient of A, and

$e_A{}^B$ = Interaction coefficient of A with respect to B

For larger number of solutes, the expression can be generalised to include all solutes.

For steel making reactions, the 'e' values have been well researched and tabulated. Using the Tables and the Wagner's expression for activity coefficient, precise thermodynamic limits for refining reactions occurring during steel making, can be established. This can also serve as guide towards process modifications so that the desired refining goal may be more closely attained.

The self interaction coefficient for hydrogen and nitrogen in liquid steel is found to be close to zero, thereby meaning that, in the region of interest they obey Henry's Law. For carbon, this value is 0.22, for oxygen, –0.20 and for sulphur, –0.028 meaning carbon exhibits positive deviation while oxygen and sulphur exhibit varying degrees of negative deviations.

7.3. EQUILIBRIUM CONSTANT

Let us consider a chemical reaction of the type

$$A + B = C + D$$

which is taking place at constant temperature and pressure. Free energy change of this reaction, ΔF, would be given by

$$\Delta F = (F_C + F_D) - (F_A + F_B)$$

If we allow sufficient time, the reaction would proceed to an equilibrium state. Let us also consider a hypothetical situation when each of the reactants and products are in standard state. In that case, the standard free energy change, $\Delta F°$, of the reaction would be

$$\Delta F° = (F_C° + F_D°) - (F_A° + F_B°)$$

On taking difference of the two reactions we get

$$\begin{aligned}\Delta F - \Delta F° &= \{(F_C - F_C°) + (F_D° + F_D°)\} - \{(F_A - F_A°) + (F_B° - F_B°)\\ &= RT \ln a_C + RT \ln a_D - RT \ln a_A - RT \ln a_B\\ &= RT \ln \{(a_C * a_D)/(a_A * a_B)\}\\ &= RT \ln k\end{aligned}$$

where,

$$k = (a_C * a_D)/(a_A * a_B)$$

As the reaction approaches equilibrium, activity values get adjusted among themselves. At equilibrium, the free energies of products and reactants become equal which means, ΔF = 0. Then the above equation reduces to

$$-\Delta F° = RT \ln k$$

or $$\Delta F^\circ = -RT \ln k$$

Since ΔF° has a definite constant value for the system, 'k' is also a constant for that system. We refer to 'k' as the 'equilibrium constant.' Readers may note that this relationship for 'k' is analogous to the classical postulates of equilibrium constant where it is defined as the ratio of the products of concentrations of reaction products and reactants. Only concentration terms have been replaced by activity terms. Here the science of thermodynamics has provided us with a correction for the classical postulate.

Illustration

Let us consider an actual chemical reaction. During sponge iron making in a rotary kiln, the main reaction is,

$$FeO + CO = Fe + CO_2$$

$$\Delta F^\circ = -RT \ln k$$

$$k = (a_{Fe} * a_{CO_2})/(a_{FeO} * a_{CO})$$

This can be approximated to

$$k = pCO_2 \,/\, pCO$$

i.e., ratios of the partial pressures of CO_2 and CO. Fe and FeO are in pure solid states and their activities can be taken to be unity.

From thermodynamic tables we get the following data for this equation,

$$\Delta F^\circ = -\,4{,}190 + 5.13T \quad \text{(calories)}$$

Temperatures involved are in the vicinity of 1000 °C (*i.e.* 1273 °K). Therefore ΔF° value is + 2,340 calories. The k value comes out to be about 0.4. This means that partial pressure of CO_2 should not be more than 0.4 times the partial pressure of CO for sponge iron making process to proceed. Stated differently, the partial pressure of CO should be at least 2.5 times that of CO_2 for successful sponge iron making. No wonder that in sponge iron making process all care is taken to keep CO_2 as far away from the sponge iron bed as possible. Further, conditions are so maintained that any CO_2 produced in the solid bed is quickly converted back to CO.

For any general temperature T (Kelvin), k value for this reaction is given by

$$\begin{aligned} -\,4190 + 5.13T &= -\,RT \ln k \\ &= -\,2.303\, RT \log k \\ &= -\,2.303 * 1{,}987 * T * \log k \\ &= -\,4.576 * T * \log k \end{aligned}$$

or, $$\log k = 915.6/T - 1.121$$

At a temperature of about 544°C (817 °K), log k is zero, *i.e.*, k is one, which means that partial pressure of CO just needs to be greater than that of CO_2 for the reduction reaction to proceed. From the feasibility point of view, this is a more favourable situation in comparison to the operating temperature of about 1000 °C in the commercially-operating rotary kilns, and about 900 °C in the commercially operating shaft furnaces. Temperatures in commercial processes are kept relatively high because the rate of reduction below 900 °C is too slow to be industrially competitive. Further, the drawback of lower feasibility is made up by maintaining a large excess of CO over CO_2 (also of H_2 over H_2O in the shaft-based process). Further, in rotary kiln process, the CO has to be generated *in situ* by the carbon in coal within the bed, by the reaction

$$CO_2 + C = 2CO$$

This coal gasification reaction attains appreciable rate for sustaining the reduction reaction only at about 1000 °C for most of the reactive chars (devolatilised coal). Those who have access to very highly reactive coals (coals which give on devolatilisation very highly reactive chars), operate their rotary kilns below 1000 °C.

7.3.1. Generalised Equilibrium Constant

If the reaction is of the type

$$xA + yB = pC + qD$$

Then, $\Delta F° = -RT\{(p.\ln a_C + q.\ln a_D) - (x.\ln a_A + y.\ln a_B)\}$

$$= RT.\ln\{(a_C^p * a_D^q)/(a_A^x * a_B^y)\}$$

Here, $k = (a_C^p * a_D^q)/(a_A^x * a_B^y\}$

If we take a more general example of a reaction, such as

$$aA + bB + cC + \cdots\cdots = pP + qQ + rR + \cdots\cdots$$

the equilibrium constant expression would become,

$$k = (a_P^p * a_Q^q * a_R^r * \cdots\cdots\cdots)/(a_A^a * a_B^b * a_C^c * \cdots\cdots\cdots)$$

7.4. ENTROPY CHANGE DURING IDEAL MIXING OF GASES

Let us consider mixing of two gases *A* and *B* to form a homogeneous mixture (a solution). Free energy changes of the constituents involved are

$$\Delta F_A = RT \ln a_A$$

$$\Delta F_B = RT \ln a_B$$

$$\Delta F_{\text{mixture}} = N_A\,\Delta F_A + N_B\,\Delta F_B$$

And if the constituents A and B behave ideally in the mixture, their activities can be replaced by atom fractions. Thus

$$\Delta F_{\text{mixture}} = RT\,(N_A \ln N_A + N_B \ln N_B)$$

But by definition,

$$\Delta F_{\text{mixture}} = \Delta H_{\text{mixture}} - T\Delta S_{\text{mixture}}$$

For ideal mixing case, $\Delta H_{mixture}$ (representing energies of interaction between atoms), should be zero, and this gives

$$\Delta S_{mixture} = -R\,(N_A \ln N_A + N_B \ln N_B)$$

Since $\ln N_A$ and $\ln N_B$ are always negative, entropy increases during formation of a homogeneous mixture, even though there has not been any heat exchange. There would not be any change anywhere else since this type of mixing takes place spontaneously and on its own, and this term indeed is the entropy increase of the universe.

We know by experience that when two dissimilar gases are kept in two different compartments and if the partition between them is removed, the gases mix spontaneously and tend to make a homogeneous mixture. On the other hand, we never find that from a mixture of gases the molecules segregate to form two volumes of pure gases spontaneously. If this were to happen, entropy of the universe would have decreased by an amount $R\,(N_A \ln N_A + N_B \ln N_B)$.

From the above relationship, we can sense that molecular distribution has something to do with entropy generation during mixing. R, the Gas Constant, has the dimensions of entropy (energy per unit absolute temperature) and N_A and N_B are dimensionless mole fractions defined by

$$N_A = n_A/(n_A + n_B) \text{ and } N_B = n_B/(n_A + n_B)$$

where n represents number of molecules. For a molar volume $n_A + n_B$ equals Avogadro's number, *i.e.*, number of molecules in a gram mole of an ideal gas, represented by N_a and having an approximate value of $6.023 * 10^{23}$.

Thus, in a system involving molar volume, $N_A = n_A/N_a$ and $N_B = n_B/N_a$

This way, during idealised mixing, the entropy generation is related to the number of molecules mixing together. This can lead one to infer that during such a mixing process (idealised mixing), the entropy generation, which is a macro property, can be linked to the number of molecules mixing together, which is a micro property. This aspect has been further discussed in Chapter 9.

7.4.1. Regular Solutions

There are not many examples of ideal solutions of interest in metallurgical processes. But a large number of solutions between two metals or two metal oxides, both solid and liquid solutions, are such that they can be termed as **half ideal**. They are termed as "Regular Solutions" and are characterised by ideal entropy of mixing even though they have a non-zero heat of mixing. In such a case, if sufficient data are not available, but there is a clue that the system could be regular, the free energy of mixing can be reasonably estimated. It is relatively easy to measure the heat of mixing calorimetrically (or by some other means) and then we can estimate the free energy of mixing to be

$$\Delta F_{mixing} = \Delta H_{mixing} + \Sigma\, RT\, N_i \ln N_i$$

Chapter-8

Measurment of Thermodynamic Properties and Some Practical Applications

As stated earlier, the most important applications of chemical and metallurgical thermodynamics are in the processes of synthesis, extraction, refining, etc. The free energy change is by far the most important of the thermodynamic properties as it is linked to identifying the limits of an actual process. The free energy in turn is dependent on other parameters like activity, interaction co-efficient, entropy, enthalpy, heat capacity, etc.

The free energy change of a reaction is best estimated from an input of enthalpy *i.e.*, heat change of a reaction and the entropy change in the reaction. The entropy change can often be estimated from the enthalpy change at equilibrium temperature. Entropy at different temperatures can be derived using heat capacity data. All these point to the fact that heat measurement is one of the more important parameters needed in the measurement or estimation of thermodynamic properties. **Calorimeter** is one of the more accurate techniques of measuring heat, whether employed for estimation of heat of reaction or used for the estimation of heat capacity. Heat change in a reaction or transformation can also be measured through the Differential Thermal Analysis (DTA) technique, which is a much simpler technique, but with a lower accuracy.

8.1. CALORIMETRY

Calorimetric technique using the "Mixture Method" has been extensively employed for estimating heat capacities of stable solid substances over a range of temperatures. A furnace, whose temperature is accurately controlled, is placed just above the mouth of the calorimeter taking care that the furnace heat does not interfere with calorimetric measurements. The solid material of interest is properly soaked in the even temperature zone of the furnace. At a suitable time it is dropped into the calorimeter where it exchanges heat under controlled conditions with the material surrounding it (generally a suitable liquid, but sometimes a highly-conducting solid as well).

Since the object of interest is dropped into the calorimeter, this type of calorimeter is often referred to as a "Drop Calorimeter." Since the object of interest exchanges heat by mixing with the calorimeter object (liquid or solid) the technique is referred to as the "Mixture

Method." The calorimeter is generally kept well insulated so that loss or gain of heat from the environment is minimum. A sensitive thermometer estimates the temperature changes inside the calorimeter. A mercury "Beckmann Thermometer" was used extensively in the past but now many reliable and more accurate devices are available such as **resistance thermometers** and also semi-conducting devices like **thermistors**.

Even though a calorimeter may be well insulated, but it does lose or gain heat during measurement. The success of experiment depends generally on the ability to accurately estimate this loss or gain and applying correction for it.

8.1.1. Solution Calorimeter

We would like to present here some of our first-hand experiences on the application of calorimeter to the measurement of thermodynamic properties. We would like to start with the measurement of heat of formation of Calcium Chromate ($CaCrO_4$) from Calcium Oxide (CaO) and Chromic Acid (CrO_3). Fig. 8.1 presents schematically the calorimeter employed for this purpose. The figure is self-explanatory. The constant temperature surrounding was maintained by keeping the calorimeter assembly immersed in a thermostatically-controlled water bath kept at 27 °C. Since the protruding shafts resemble somewhat the shaft of a submarine and since major part of the calorimeter is kept immersed in water, this type of calorimeter is sometimes referred to as the "Submarine Type Calorimeter." Further, a calorimeter kept within a constant temperature surrounding, is referred to as an "Isoperibol Calorimeter;" *peribol* is a Greek word meaning surroundings.

In separate experiments, each of the reactants and products (*viz.*, CaO, CrO_3 and $CaCrO_4$) were dissolved in 6N hydrochloric acid (HCl) kept in the Dewar flask, which is the main container of the calorimeter. Since CaO is unstable in normal atmosphere, it had to be prepared and sealed in a thin wall glass bulb within a "Dry Box" where moisture and CO_2-free atmosphere is maintained. For maintaining uniformity, CrO_3 and $CaCrO_4$ were also sealed in thin-walled glass bulbs. The sealed and weighed samples were attached to a glass rod, the other end of which protruded out of the top of calorimeter in such a way that it could be manually rotated. The sample containing bulbs were fixed in such a way that on rotating the glass rod by 90°, the bulb could be struck against the wall of the calorimeter and broken so that the sample could react with the calorimeter liquid (6N HCl).

The temperature changes were measured using the thermistor located inside the calorimeter and its leads were taken out and connected to a Wheatstone Bridge comprising of "Decade Resistance Boxes" (fine and accurate set of electrical resistances) and a moving coil galvanometer.

A continuously operating mechanical stirrer ensured quick mixing of material inside the calorimeter. Similar to the location of the thermistor, a calibrating resistance coil protected with glass was kept inside the calorimeter for measuring the water equivalent of the calorimeter. Calibrated current was passed for specified time and the temperature rise was monitored for calculating the water equivalent.

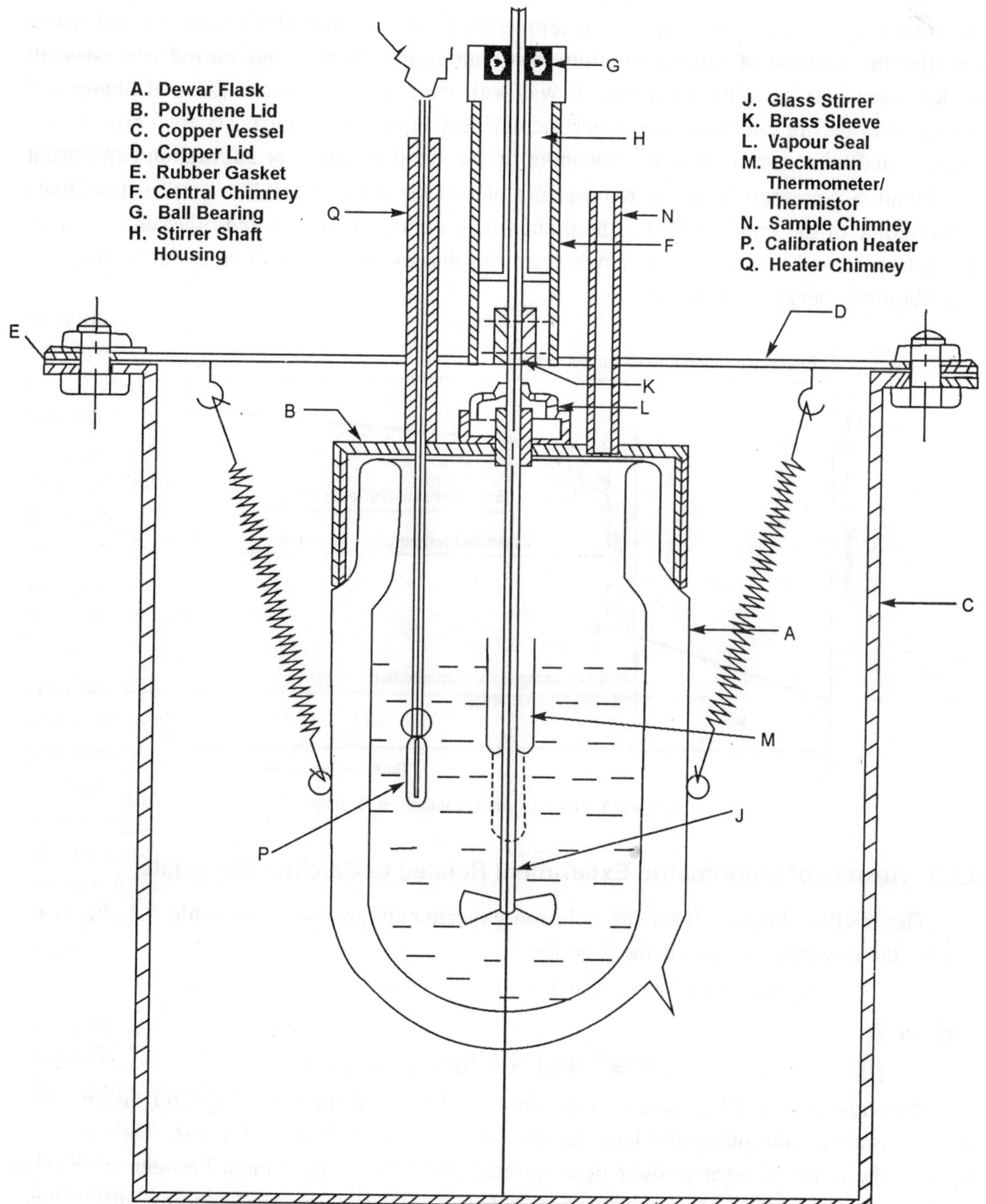

Figure 8.1 Acid solution calorimeter

To correct for radiation loss and the continuous energy input through the continuously operating mechanical stirrer, change in temperature was monitored for some period before and after the addition of sample or addition of calibrating energy. This period was normally not less than half an hour. Generally it was well over an hour. These rates of changes of temperature, before and after the experimental heat input, gave the basis for correction of radiation and other contributors of heat input during the dissolution or calibration experiment.

Figure 8.2 presents a typical temperature plot of the calorimeter. The vertical line drawn in the middle of the graph indicates the point where the radiation loss and other energy inputs are neutralised and thus the exact temperature rise due to sample addition or due to the input of calibrating energy is obtained.

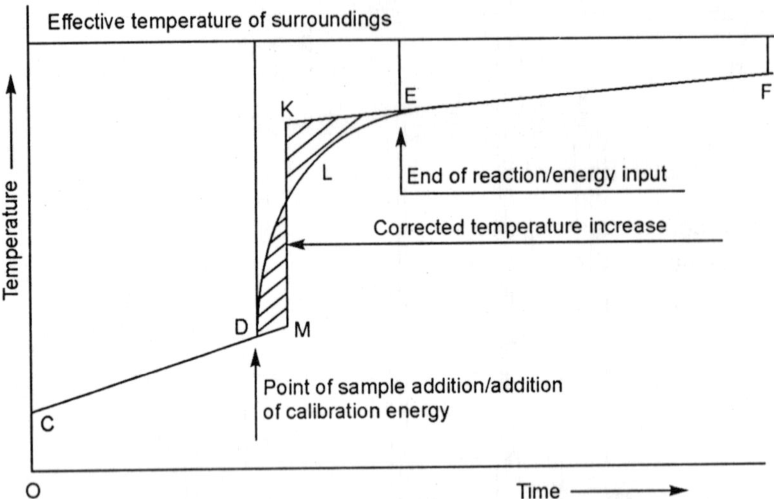

Figure 8.2. The integration method of Rossini

8.1.2. Results of Calorimetric Experiment Related to Calcium Chromate

The results obtained from the solution experiments are given in Table 8.1. From the results, the enthalpy change of the reaction

$$CaO(s) + CrO_3(s) = CaCrO_4(s)$$

is given by

$$\Delta H = -44{,}435 \pm 200 \text{ cals. at } 27\ °C$$

Wartenberg et al. (1937) have determined the heat of formation of $CaCrO_4$ using a heat of neutralisation calorimeter and have reported the value – 49.75 *kCal* for ΔH. They have not reported the limits of error in their measurement. But there is agreement between their value and the value reported in the present work (*Prasad & Abraham, 1970*) taking into account the error of ±3000 calories, estimated for the results of Wartenberg. Subsequently, *Jacob et al.* (*1992*) re-estimated the value by EMF method and reported a value of 47.68 ± 1.2 kCal.

Table 8.1 Heats of Solution in 6N HCl at 27 °C.

Substance	*Heat evolved (-ΔH solution), (calories)*	*Mean ΔH solution, (calories)*
CaO	48,223 48,022 48,033	– 48,093 ± 130
CrO_3	691 523 779	– 664 ± 150
$CaCrO_4$	4,282 4,361 4,319	– 4,321 ± 40

8.2. ACTIVITY MEASUREMENTS

The most common technique of measuring activities of oxide constituents (*e.g.*, FeO in solid solution with MnO) has been through equilibration with suitable gas mixtures. In case of FeO-MnO, solid solutions, *Foster and Welch (1956)* equilibrated them with H_2/H_2O gas mixtures in the temperature range 850 to 1150 °C.

$$(FeO) + H_2(g) = Fe + H_2O(g)$$

$$\Delta F° = -RT \ln k$$

$$= -RT \ln [(a_{Fe} * a_{H_2O})/(a_{FeO} * a_{H_2})]$$

$$= -4.576T \log[(p_{H_2O}/p_{H_2})*(1/a_{FeO})]$$

where, (FeO) represents FeO dissolved in FeO- MnO solid solution. FeO and MnO are completely miscible in the temperature range considered and both these constituents and their solid solutions have NaCl type "Face Centred Cubic" or FCC crystal lattice structure.

Thus,

$$a_{FeO} = e^{(\Delta F°/4.576T)}/(pH_2/pH_2O)$$

and by substituting the values of equilibrated p_{H_2}/p_{H_2O} ratio, absolute temperature and standard free energy change of reaction, the activity values were computed. Their results indicated ideal mixing behaviour in this system.

Schenck et al. (1929) had earlier used the same technique but used CO/CO_2 gas mixture for equilibration. They had reported moderate positive deviation from ideality, which was also confirmed by *Swerdtfeger and Muan (1967)*. However, *Engell (1962)* reported a much higher positive deviation using the solid electrolyte galvanic cell technique, about which we would discuss in the next section.

The main uncertainty in the gas equilibration technique has been to judge accurately whether equilibrium has really been achieved or not. Consequently, the extent of range of error in such measurements is also difficult to estimate. It was, therefore, decided to check and re-estimate the reported data with independent techniques.

8.2.1. The High Temperature Solid Electrolyte Galvanic Cell Technique

When dissimilar metals or electrodes are immersed in an electrolytic solution with common ions, an electromotive force (EMF) develops between the electrodes. This is the principle behind formation and working of galvanic cells. The EMF is characteristic of the free energy change in ion exchange (i.e. the cell reaction).

$$\Delta F = -n E F_a$$

where ΔF = Free energy change for the cell reaction

n = number of Faradays to be passed across the cell to effect the cell reaction

E = Electromotive force developed in the cell

F_a = Faraday Constant

The free energy of reaction can be related to activities. One condition here is that the electrolyte should not conduct electricity through the flow of free electrons (as in the case of metallic conductors) and the electricity transport is entirely through transport of ions.

Solid oxides and fluorides satisfy the same conditions, and when used as electrolyte for forming solid state galvanic cells they are referred to as "Solid Electrolytes."

These solid electrolytes have a very high resistance at room and moderate temperatures. Therefore, they are suitable for forming galvanic cells only at high temperatures, generally between 850 and 1150 °C.

Engell (1962) studied the activities of FeO in FeO-MnO solid solutions, by setting up the cell

$$\text{Pt} \mid \text{Fe, FeO} \parallel ZrO_2 - \text{CaO} \parallel \text{(Fe, Mn)O, Fe} \parallel \text{Pt}$$

Here (Fe, Mn)O represents the solid solution between FeO and MnO. An inert atmosphere was required to be maintained to prevent oxidation of Fe (metal) and Fe^{2+}. Zirconia, ZrO_2 is a solid electrolyte, whose capacity to transport O^{2-} ions is enhanced by doping it with lime, CaO. This doping of Ca^{2+} ions in Zr^{4+} matrix leads to creation of vacant spaces in the lattice structure. These vacant spaces enable O^{2+} ions to move more freely than it was otherwise possible.

Here the cell reaction is

$$\text{FeO} = (\text{FeO})_{\text{MnO}}$$

where, $(\text{FeO})_{\text{MnO}}$ represents solid solution of FeO in MnO.

From free energy considerations we get

$$2E F_a = RT \ln a_{\text{FeO}}$$

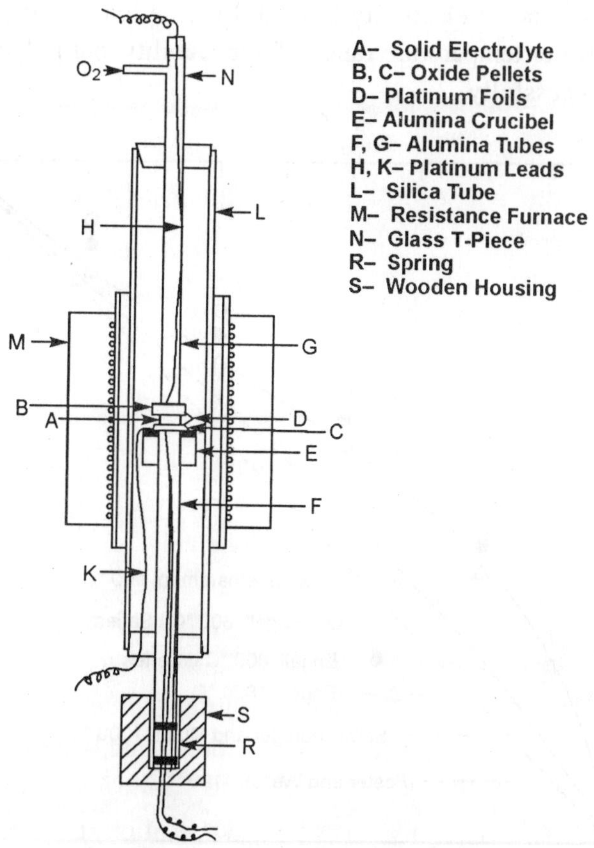

Figure 8.3 Solid electrolyte cell assembley

By measuring the equilibrium EMF of this cell, *Engell (1962)* reported high positive deviation from ideality in this system. *Seetharaman & Abraham (1968)* used a similar technique but used Thoria doped with Yttria as the solid electrolyte. They used the following cell

$$Pt \mid Fe + (Fe, Mn)O \parallel ThO_2 - Y_2O_3 \parallel Ni + NiO \mid Pt$$

Their results were in rough agreement with the values reported by *Engell (1962)*. The general arrangement of solid electrolyte cell assembly is presented in Fig. 8.3. Here, flow of oxygen through the cell is indicated, which is required in some types of cells, as we shall see later in the case of $CaCrO_4$. In the preceding examples, oxygen-free inert gas were required and for that only the gas flow system had to be modified. The activity results for the FeO-MnO system are presented in Fig. 8.4.

8.2.2. Fluoride Solid Eectrolytes

The fluorides of calcium, magnesium, lead and a few other metal fluorides, being strongly ionic compounds, have served as solid electrolytes for the purpose of activity measurements.

These fluorides naturally conduct electricity through fluoride ion (F^-) transport. Single crystals of these fluorides have wider temperature range of applicability, but polycrystalline electrolytes have also been used successfully.

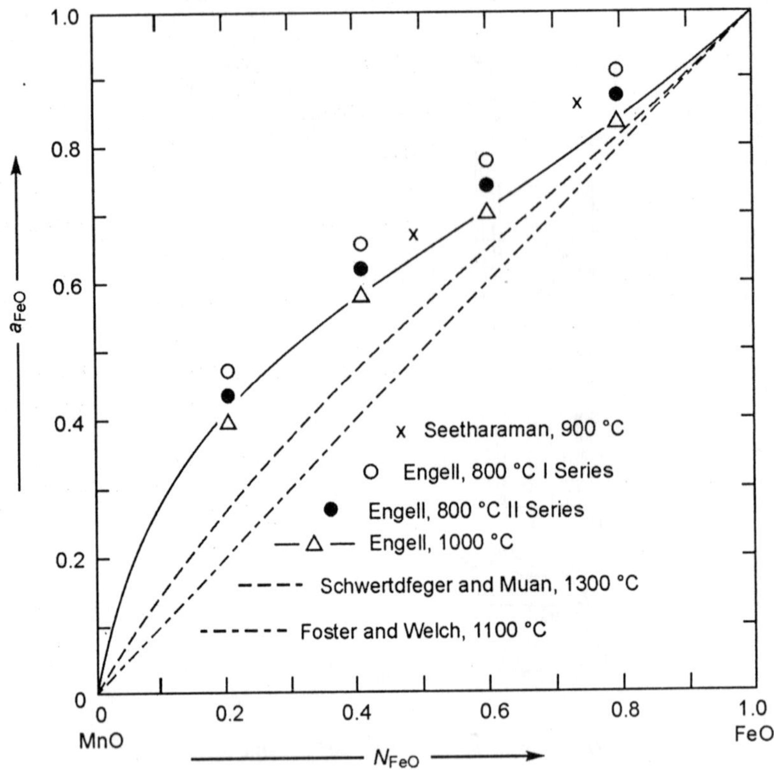

Figure 8.4 Activities of FeO in the system FeO-MnO

8.2.3. Galvanic Cells with CaF_2 Solid Electrolyte

CaF_2 solid electrolyte has been used to measure the free energy of formation of Calcium Vandates, $CaO.V_2O_5$, $2CaO.V_2O_5$ and $3CaO.V_2O_5$ by measuring the electromotive forces of the following cells (*Prasad & Abraham, 1971*)

$Pt,O_2 \mid CaO \parallel CaF_2 \parallel 3CaO.V_2O_5 + 2CaO.V_2O_5 \mid Pt,O_2$

$Pt,O_2 \mid 3CaO.V_2O_5 + 2CaO.V_2O_5 \parallel CaF_2 \parallel 2CaO.V_2O_5 + CaO.V_2O_5 \mid Pt,O_2$

$Pt,O_2 \mid 2CaO.V_2O_5 + CaO.V_2O_5 \parallel CaF_2 \parallel CaO.V_2O_5 + V_2O_5 \mid Pt,O_2$

$Pt,O_2 \mid CaO + 2CaO.V_2O_5 \parallel CaF_2 \parallel 2CaO.V_2O_5 + CaO.V_2O_5 \mid Pt,O_2$

and

$Pt,O_2 \mid 3CaO.V_2O_5 + 2CaO.V_2O_5 \parallel CaF_2 \parallel CaO.V_2O_5 + V_2O_5 \mid Pt,O_2$

Dry and CO_2 free atmosphere was maintained during the measurements by passing purified oxygen gas continuously. Results of measurements are presented in Table 8.2.

Table 8.2. Free energy values calculated from e.m.f. measurements for the system $CaO\text{-}V_2O_5$

A. Measured values

Cell reaction	*Temperature range of measurement (°C)*	*EMF, mV*	*ΔF (= –2EF$_a$) (calories)*
$CaO + 2CaO \, . \, V_2O_5 = 3CaO \, .V_2O_5$	650 – 750	517 ± 5	–23,850 ± 230
$3CaO \, . \, V_2O_5 + CaO \, . \, V_2O_5 = 2(2CaO \, . \, V_2O_5)$	500 – 550	73 ± 5	– 3,370 ± 230
$2CaO \, . \, V_2O_5 + V_2O_5 = 2(CaO \, . \, V_2O_5)$	500 – 550	190 ± 5	– 8,770 ± 230
$CaO + CaO \, . \, V_2O_5 = 2CaO \, . \, V_2O_5$	650 – 550	596 ± 5	– 27,495 ± 230
$3CaO \, . \, V_2O_5 + V_2O_5 = 2CaO \, . \, V_2O_5 + CaO \, . \, V_2O_5$	550	259 ± 5	– 11,950 ± 230

B. Calculated from the above experimental values

Reaction	*ΔF (500 – 700 °C) calories*
$CaO + V_2O_5 = CaO \, .V_2O_5$	– 35,990 ± 400
$2CaO + V_2O_5 = 2CaO \, .V_2O_5$	– 63,210 ± 690
$3CaO + V_2O_5 = 3CaO \, .V_2O_5$	– 87,060 ± 860

CaF_2 solid electrolyte has also been used for measuring free energy of formation of calcium chromate by measuring the EMF of the following cell (*Prasad & Abraham, 1970*)

$Pt,O_2 \mid CaO \parallel CaF_2 \parallel Cr_2O_3 + CaCrO_4 \mid Pt,O_2$

Here the cell reaction is

$$CaO + 1/2 \; Cr_2O_3 + 3/4 \; O_2 = CaCrO_4$$

Since oxygen is an essential part of this cell reaction, maintaining an atmosphere of pure and dry oxygen is essential for obtaining a stable cell EMF. The results of these measurements are presented in Table 8.3.

Table 8.3 Free energy values calculated from the e.m.f. measurement of the cell involving $CaCrO_4$

Cell assembly			
$Pt, O_2(g)$ I $CaO(s)$ II $CaF_2(s)$ II $CaCrO_4(s) + Cr_2O_3(s)$ I $Pt, O_2(g)$			
Cell reaction			
$CaO + 1/2Cr_2O_3 + 3/4\ O_2 = CaCrO_4$			
Temperature		*EMF (mV)*	*ΔF (Calories)*
°C	°K		
600	873	525 ± 10	– 24,215 ± 460
700	973	480 ± 10	– 22,140 ± 460
800	1073	437 ± 2	– 20,110 ± 90

Other than compounds, the activities of CaO in solid solutions have also been measured using CaF_2 solid electrolyte. The CaO-CdO system is one such example. Here also CaO and CdO are both having NaCl-type FCC lattice structure and are fully miscible in the entire composition range. Through the measurement of EMF of the following cell, the activities of CaO in the CaO-CdO solid solutions were determined (*Prasad et al., 1975*)

$$Pt, O_2 \mid CaO \parallel CaF_2 \parallel (Ca,Cd)O \mid Pt, O_2$$

They also checked the compatibility of activities with heats of mixing through calorimetric measurements. Both activities and enthalpies of mixing indicated ideal mixing in this system. Results are presented in Table 8.4 and Fig. 8.5.

Table 8.4. Measured values of activities of CaO and heats of mixing in CaO-CdO solid solutions.

Cell assembly		
$Pt, O_2(g)$ I $CaO(s)$ II $CaF_2(s)$ II $(Ca,Cd)O$ I $Pt, O_2(g)$		
Typical EMF data obtained with 50:50 CaO-CdO solid solution		
Temperature (°C)	*EMF (mV)*	a_{CaO}
800	35 ± 2	0.48 ± 0.03
840	36	0.47
850	36 ± 0.5	0.48
870	37 ± 1	0.47 ± 0.01

(Contd...)

Activities of CaO in the system CaO-CdO

Temp. (°C)	*Mole fraction CaO*				
	0.25	0.39	0.5	0.6	0.75
550	0.24				
660	0.255				
750		0.395		0.595	
800		0.42	0.48		
850		0.385	0.47	0.615	
935			0.58		0.757
970		0.42	0.487	0.629	0.741
1005		0.387	0.485	0.604	0.776
1030		0.394	0.509	0.585	0.734

Heats of mixing of CaO-CdO solid solution

Mole fraction CdO	*Heats of mixing (calories/mole)*	*Mole fraction CdO*	*Heats of mixing (calories/mole)*
0.2000	20 – 196 – 212	0.5000	57
0.3349	– 58	0.6101	207 6 – 130 – 146
0.4000	156 3 –107	.7500	– 10 – 119

8.2.4. Galvanic Cells with MgF_2 Solid Electrolytes

MgF_2 solid electrolytes have also been used to measure activities of MgO in both compounds (Magnesium Titanates – *Shah et al., 1970*) and solid solutions involving MgO. In the latter category, MgO-ZnO, MgO-CoO and MgO-NiO have been studies by *Raghavan et al. (1975*) and *Prasad (1971). Seetharaman (1970)* has rechecked some of the results. He found very good agreement in results obtained from the different techniques.

8.2.5. Gibbs Duhem Integration

In a system like MgO-NiO, the activities of MgO in the entire composition range were measured by using MgF_2 solid electrolyte galvanic cell, while NiO activities were separately determined using the oxide solid electrolyte galvanic cell. But mathematically, in such a miscible system, for any state property, the partial molar property of one component automatically fixes the partial molar property of the other component. Thus, if the partial molar free energy (a measure of the activity) of one component is known over the entire composition range, the corresponding partial molar free energy of the other component can be mathematically calculated.

The process of calculating the partial molar property of one component from the knowledge of partial molar property of the other component in a binary system, where the integral molar property changes continuously with the composition, is referred to as the Gibbs Duhem Integration.

Let us consider a state property G of a system of one mole consisting of N_A mole fraction of constituent A and N_B mole fraction of constituent B. Let G_A and G_B be the corresponding partial molar properties of A and B in the mixture or solution. Then by definition

$$G = N_A G_A + N_B G_B$$

and $$N_A + N_B = 1$$

In case of an infinitesimal change in composition, G will change as per the following

$$\begin{aligned} dG &= G_A\, dN_A + G_B\, dN_B \\ &= G_A\, dN_A + G_B\, d\,(1 - N_A) \\ &= G_A\, dN_A - G_B\, dN_A \end{aligned}$$

or, $$dG = (G_A - G_B)\, dN_A$$

The integral expression can be written as

$$\begin{aligned} G &= N_A\, G_A + (1 - N_A)\, G_B \\ &= G_B + N_A\, (G_A - G_B) \end{aligned}$$

On differentiating this expression we get

$$dG = dG_B + N_A(dG_A - dG_B) + (G_A - G_B)dN_A$$

Substituting the previous expression for dG we get

$$(G_A - G_B)\, dN_A = dG_B + N_A(dG_A - dG_B) + (G_A - G_B)\, dN_A$$

or $$dG_B + N_A(dG_A - dG_B) = 0$$

or $$dG_B\, (1 - N_A) = -\, N_A\, dG_A$$

or $$\mathrm{d}G_B = -\, (N_A/(1 - N_A))\, dG_A$$

Integrating, we get

$$G_B = -\int (N_A/(1 - N_A))\, dG_A$$

for N_A ranging from 0 to the given value.

The right hand expression can be graphically integrated to obtain the value of G_B. This expression is the area under the curve obtained by plotting $N_A/(1 - N_A)$ against G_A. The calculation process can best be illustrated through the following estimation of the activity of MnO from the activity values of FeO presented in Fig. 8.4.

In case of a solution, like the FeO-MnO solid solution, the integral molar free energy of mixing would be related to the partial molar values, as per their definition, in the following manner:

$$\Delta F_{\text{mixing}} = N_A \Delta\overline{F}_A + N_B \Delta\overline{F}_B$$

and $$\Delta\overline{F}_A = RT \ln a_A$$

also $$\Delta\overline{F}_B = RT \ln a_B$$

Differentiating we get

$$d\Delta F = \Delta\overline{F}_A dN_A + N_A\, d\Delta\overline{F}_A + \Delta\overline{F}_B\, dN_B + N_B\, d\Delta\overline{F}_B$$

Since ΔF_{mixing} is a state property

$$d\Delta F = \Delta\overline{F}_A dN_A + \Delta\overline{F}_B\, dN_B$$

Combining the two equations we get

$$N_A\, d\Delta\overline{F}_A + N_B\, d\Delta\overline{F}_B = 0$$

or $$RT\, N_A\, d \ln a_A + RT\, N_B\, d \ln a_B = 0$$

Hence $$d \ln a_B = -(N_A/N_B)\, d \ln a_A$$

and $$\ln a_B = -\int (N_A/N_B) d \ln a_A$$

integrated within the limits $N_A = 0$ to the required value of N_A.

For evaluating this integration, it is more convenient to use γ *i.e.*, activity coefficient terms, rather than simple activity. In γ terms, the above equation reduces to

$$d \ln (\gamma_B N_B) = -(N_A/N_B)\, d \ln (\gamma_A N_A)$$

or, $$d \ln \gamma_B + d \ln N_B = -(N_A/N_B)(d \ln \gamma_A + d \ln N_A)$$

or, $$d \ln \gamma_B = -(N_A/N_B)\, d \ln \gamma_A - (N_A/N_B)\, d \ln N_A - d \ln N_B$$

or, $$\ln \gamma_B = -\int [(N_A/N_B)\, d \ln \gamma_A - (N_A/N_B)(1/N_A)\, d N_A - (1/N_B)\, d N_B]$$

integrated between the limits $N_A = 0$ to the required value of N_A.

As $dN_A + dN_B = 0$,

Hence, $$\ln \gamma_B = -\int (N_A/N_B)\, d \ln \gamma_A$$

integrated between the same limits.

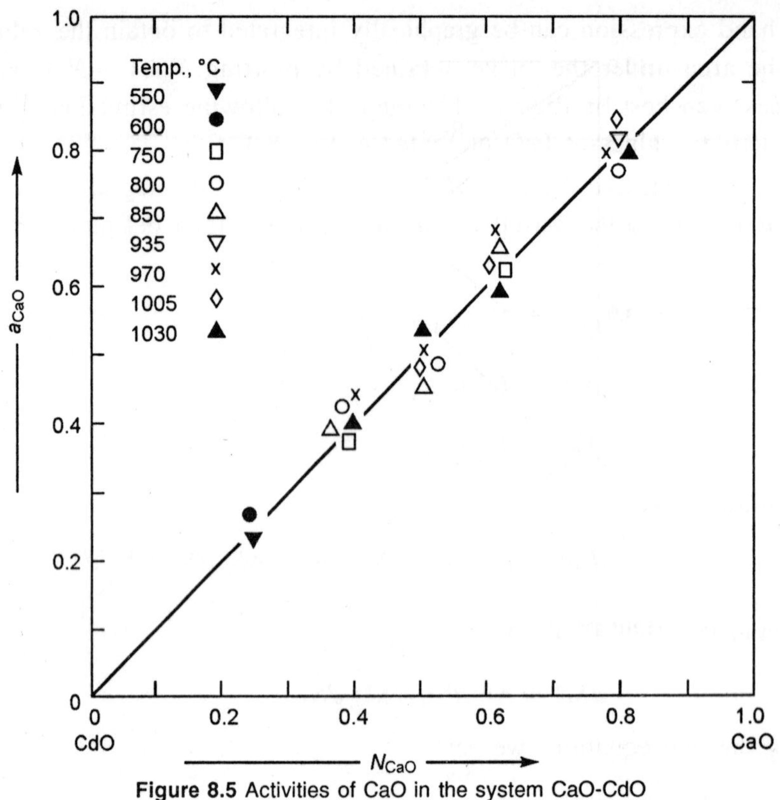

Figure 8.5 Activities of CaO in the system CaO-CdO

Accordingly, the ratio N_{FeO}/N_{MnO} has been plotted against $\log_{10}\gamma_{FeO}$ in Fig.8.6. By measuring areas under the curve in different ranges, log of activity coefficients of MnO, and thereby its activities have been estimated. The latter have been plotted in Fig. 8.7.

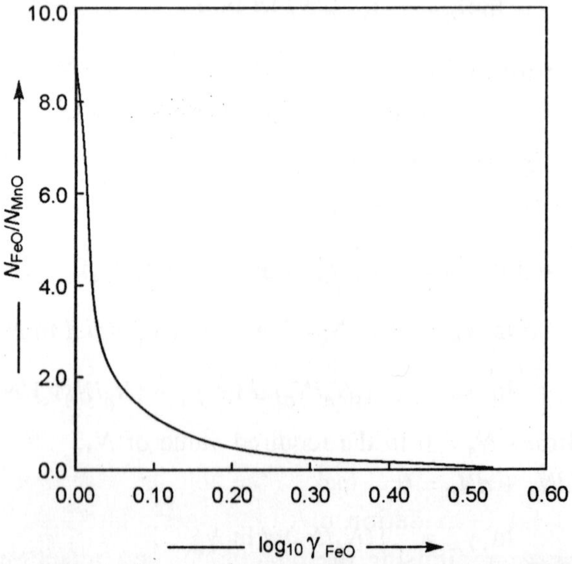

Figure 8.6. Plot for graphical estimation of $\log_{10}\gamma_{MnO}$

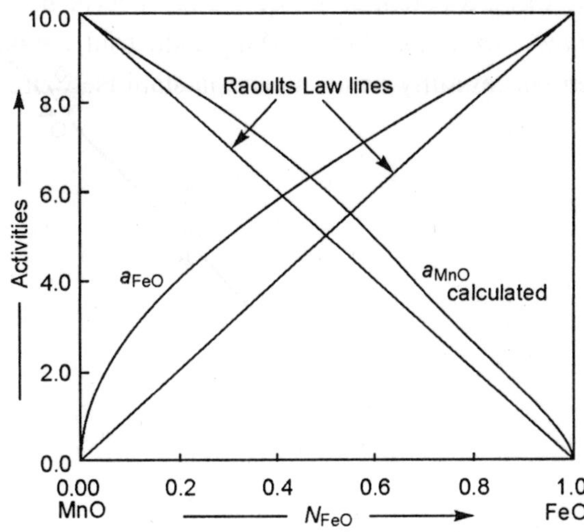

Figure 8.7 Estimation of activities of MnO in FeO-MnO system by Gibbs Duhem integration

8.2.6. Use of Solid State Galvanic Cells in Metal Extraction and Refining

Steel making is a controlled oxidation process. With the advent of oxygen lancing, refining times have become very short. It is now necessary to quickly assess whether the desired oxidation level, or the oxygen activity, has been attained in the liquid steel bath. Disposable oxygen probes utilising the high-temperature solid electrolyte galvanic cells have come in handy and are now widely used to assess the completion of oxygen blowing in the steel refining process.

Rourkela Steel Plant of Steel Authority of India Limited (SAIL) introduced the use of "Celox" oxygen sensors in 1984. Celox sensors made by M/s Electronite Corporation of Belgium was selected in view of its superior past performances as compared to other commercially available sensor like FEA probe of University of Pittsburgh, USA; Oxytip, IRSID, France and Oxypac made by ASEA, Sweden.

The Celox cell consists of a closed end solid electrolyte tube made of zirconia stabilised with magnesium oxide (MgO). Cr + Cr_2O_3 is used as the **reference electrode**. Pt/Pt-Rh(10%) thermocouple is used for *in situ* temperature measurement and the Pt lead also serves as the lead for the cell for EMF sensing. The lance is dipped into liquid steel for measuring oxygen potential. Liquid steel acts as the second electrode of the cell. As mentioned earlier, these are disposable type and can be used only once.

Celox sensors are now being used regularly in almost all the Indian steel plants. By determining the oxygen activity the steel makers are able to judge more accurately the quantity of deoxidisers required for deoxidation of liquid steel. This has helped in reducing the production cost of steel by minimising the quantity of deoxidisers and also improving the quality of steel produced through maintaining proper refining regime during steel making.

Probes for measuring oxygen potential in liquid copper have been successfully developed and used in laboratory scale (*Abraham, 1967*). Many industrial copper smelting and refining units in Europe have been successfully using magnesia-stabilised zirconia probes using air as the reference electrode.

Chapter- 9

Statistical Thermodynamics: An Introduction

Development of the subject of thermodynamics, as the readers might have noted, has been based on facts, some commonly observed facts, while others have been experimentally determined findings. These observations of facts have been at macrolevels, and have been summarised into the different laws of Thermodynamics. While the applicability of these laws at micro level can be a matter of debate, but logically the thermodynamic properties of the system determined at macro level, should be derivable from the properties of particles constituting the system. The method and principle behind this derivation forms the basis of Statistical Thermodynamics – a special branch of thermodynamics.

Any system above zero absolute temperature has particles – atoms and molecules – in constant motion. Atoms and molecules in a gaseous system possess the maximum variety of motion – transport, spin, vibratory, etc. These particles constantly interact with each other and at each interaction, quantum of motion of all these types change – and in a very random manner as number of particles involved are huge (about $6 * 10^{23}$ in a g.mole). If such interactions are all mechanical interaction (*i.e.*, free from chemical or similar changes, such as particles getting associated, or breaking up during interactions), the system remains amenable to statistical interpretation of thermodynamic properties.

9.1. BERNOULLI'S INTERPRETATION

One of the earliest attempts of interpreting properties of systems based on states of atoms or molecules constituting the system, was made by Bernoulli in 1738. He considered an ideal gas within a cylinder and a piston similar to what is presented in Fig. 4.1. He considered the force exerted by the piston to be balanced by the pressure of the gas and he considered the pressure to be generated by the impact of gas molecules on piston and the consequent momentum change. He reasoned that when volume decreases, the average distance between molecules would decrease as the cube root of volume ratio, and the number of impact on the piston on this account would increase by the same ratio. On the other hand the number of particles adjacent to the piston will increase as per the area ratio which can be expressed as the square of the cube root of volume ratio. Combining the two factors together he concluded that pressure varies as per the following:

$$P \propto V^{1/3} * V^{2/3}$$

This exactly is the statement of Boyle's Law *i.e.*,

$$P.V = \text{Constant}$$

9.2. INTERPRETATION OF ENTROPY

Statistical Thermodynamic treatment of a system is based on the simple presumption that the system is made of smaller parts, the properties of the whole can be deduced from the properties of the smaller parts. This statement can be conceived and accepted very easily. But what are these smaller parts. By early 1900 the concepts of particle nature of matter and wave nature of energy (electromagnetic radiation) were well established. But soon after it was found that energy transfer takes place in 'quantas' or particle like manner. Thus individual particles – atoms or molecules – within a system exist possessing different energy levels. Number of these energy levels in a real system are too large to conceive, but the particles can be conceived to be existing at any instant in 'groups' – each group having particles in the same energy level or the same micro-state. After each interaction between a pair of particles, momentum is exchanged, and the interacting particles change their microstate and pass onto different microstates. The overall property of the system would be the weighted sum total of the properties of the microstates. Both the particles, and the microstates they exist in, are so large in real systems that the property of the system would be governed almost entirely by the property of the most populous microstate, which would, in turn, point to the "most probable macrostate."

When dealing with a very large number of molecules the probability of any distribution other than the most probable and populous becomes very small. The molecules may then be regarded as being distributed according to the most probable distribution.

The plot of the population of microstates against energy levels of the microstates would give an inverted bell shaped plot (normal distribution; Fig. 9.1). The difference here is when total number of molecules increase, width of bell becomes thinner and thinner. At extremely large number of molecules (of the order of N_a or $6*10^{23}$) the width becomes so thin that it becomes almost a vertical line.

We may be tempted to call the group of particles existing in a microstate as a subsystem. But the word 'system' has such a connotation which has led scientists to use, instead, a term 'Ensemble' basically meaning a group. In classical thermodynamics, we say that (as per zeroeth law) two systems in thermal equilibrium with a third system are in thermal equilibrium with each other; and the same can be attributed to subsystems, but not to 'ensembles'.

Let the energy levels of the molecules be denoted by the symbols e_1, e_2, e_3, etc. and these levels are occupied by n_1, n_2, n_3, ... molecules such that the total number of molecules equal N_a (we are considering molar volume of an ideal gas). If the total energy of the system be U then

$$\sum \varepsilon_i n_i = U$$

and $$\sum n_i = N_a$$

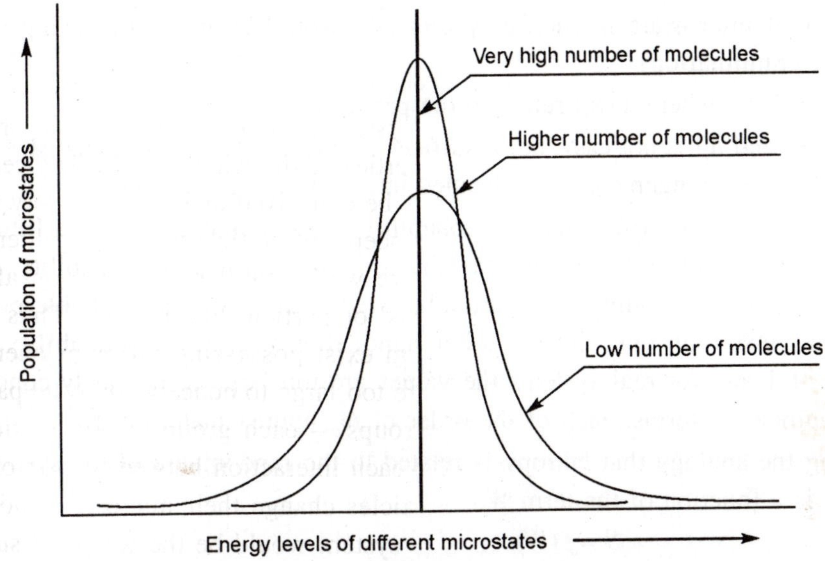

Figure 9.1 Energy level distribution chart of an ideal gas
(shows that macro properties are governed by the most populous micro-state)

At constant temperature and pressure, U would have a fixed value. Thus, even though the values of individual n_i's are changing with time (*i.e.*, changing with each collision), the sum total of energies of individual molecules remain constant. The above two equations are in fact boundary conditions imposed on n_i and $\varepsilon_i n_i$.

This normal distribution of population of molecules is represented, as per classical mechanics, by the following equation

$$n_i = n_o * e^{(-\varepsilon_i / kT)}$$

where n_o is the number of molecules in the lowest energy level and T is the absolute temperature of the system.

This distribution equation is known as the Maxwell-Boltzman distribution.

It can be further shown by classical mechanics that

$$n_o = N_a / \sum e^{(-\varepsilon_i / kT)} = N_a / Q$$

where $$Q = \sum e^{(-\varepsilon_i / kT)}$$

Q is referred to as the "Partition Function." All thermodynamic functions can be expressed as functions of Q. Evaluation of Q thus makes it possible to calculate Gibbs & Helmohltz free energies, entropy, internal energy, etc.

9.3. THERMODYNAMIC PROBABILITY

The number of microstate in such a system (W) would be given by (using method of permutation and combination)

$W = n! \,/\, \Pi n_i!$ where $\Pi n_i!$ refers to the product of all $n_i!$

Conceiving $n!$ and $n_i!$ is not easy, but it suffices at this point to assume that W (number of microstate) deals with arrangement of molecules into different microstates.

W is referred to as "Thermodynamic Probability." This nomenclature is a little difficult to digest, since in Probability Theory we say that maximum possible probability value is 1 (one), which refers to the certainty of a particular event. All other less probable events have probability values less than one. On the other hand, thermodynamic probability, W has a minimum value of 1 and for real systems the values are too large to be easily conceivable – multiples of a number of terms, each of the order of N_a, approximately $6*10^{23}$.

By applying the analogy that entropy is related to the randomness of the system we can say that entropy is a function of the term W.

$$S = f(W)$$

By applying the principle of maximisation of entropy in mixing, and extended mathematical derivation, the following differential equation can be derived

$$W.d^2S/dW^2 + dS/dW = 0$$

which leads to

$$S = C_1 \ln W + C_2$$

where C_1 and C_2 are integration constants.

Plank postulated that the second constant is zero and later showed that C_1 is same as k. The relation thus became

$$S = k \ln W$$

where k is called the Boltzman Constant and is related to Gas Constant and Avogadro's Number as given below

$$k = R/N_a$$

Since,

$$W = n!/\Pi\ n_i!$$

and

$$\ln W = \ln n! - \Sigma\ \ln n_i!$$

Using Stirling's approximation, *i.e.*,

$$\ln n! = n \ln n - n \text{ (for large } n)$$

we get,

$$\ln W = (n \ln n - n) - \Sigma (n_i \ln n_i - n_i)$$

For one mole of ideal gas

$$n = N_a, \text{ and}$$

$$\ln W = (N_a \ln N_a - N_a) - \Sigma (n_i \ln n_i - n_i)$$

$$= N_a \ln N_a - \Sigma\ n_i \ln n_I \quad \text{since } \Sigma\ n_i = N_a$$

Hence, $$S = k (N_a \ln N_a - \Sigma n_i \ln n_i)$$

9.4. IDEAL MIXING OF GASES

Let us consider ideal mixing of two gases A and B. For convenience we are taking molecules of A (n_A) and B (n_B) in such a way that on mixing a molar volume is obtained; *i.e.* $n_A + n_B = N_a$.

For gas A

$$W_A = n_A!/\Pi\ n_i!$$

and $$S_A = k \ln W_A$$

$$= k (\ln n_A! - \Sigma \ln n_i !)$$

$$= k \{(n_A \ln n_A - n_A) - \Sigma (n_i \ln n_i - n_i)\}$$

$$= k (n_A \ln n_A - n_A - \Sigma n_i \ln n_i + \Sigma n_i)$$

But $$\Sigma n_i = n_A$$

Hence, $$S_A = k (N_a \ln n_a - \Sigma n_i \ln n_i)$$

Similarly for gas B

$$W_B = n_B!/\Pi\ n_j!$$

and $$S_B = k (n_B \ln n_B - \Sigma n_j \ln n_j)$$

The possible energy states of molecules of B are different from the possible energy states of molecules of A, since the species are different and the molecules are able to take up different energy configurations.

We have, therefore, used a different subscript $_j$ to denote these different possible energy levels.

On mixing the gases we get 1 mole of the mixture containing N_a molecules. If the mixing takes place ideally *i.e.,* if mixing is random and without any association of molecules, we get, for the mixture, by method of Permutation and Combination,

$$W_{\text{mixture}} = N_a! / (\Pi n_i! * \Pi n_j!)$$

and $$S_{\text{mixture}} = k (N_a \ln N_a - \Sigma n_i \ln n_i - \Sigma n_j \ln n_j)$$

Hence, $\Delta S_{\text{mixture}} = k\,[(N_a \ln N_a - n_A \ln n_A - n_B \ln n_B)$

$$- (\sum n_i \ln n_i + \sum n_j \ln n_j - \sum n_i \ln n_i - \sum n_j \ln n_j)]$$

$$= k\,(N_a \ln N_a - n_A \ln n_A - n_B \ln n_B)$$

If we denote mole fraction of A and B in the mixture as N_A and N_B,

$$N_A = n_A/N_a \quad i.e.,\ n_A = N_A * N_a$$

Similarly $n_B = N_B * N_a$ and $N_A + N_B = 1$

Hence, $\Delta S_{\text{mixture}} = k\,(N_a \ln N_a - N_A N_a \ln N_A N_a - N_B N_a \ln N_B N_a)$

$$= k\,(N_a \ln N_a - N_A N_a \ln N_A - N_A N_a \ln N_a - N_B N_a \ln N_B - N_B N_a \ln N_a)$$

$$= k.N_a \{ \ln N_a - (N_A \ln N_a + N_B \ln N_a) - (N_A \ln N_A + N_B \ln N_B)\}$$

$$= k.N_a \{\ln N_a - (\ln N_a)(N_A + N_B) - (N_A \ln N_A + N_B \ln N_B)\}$$

$$= R\,(\ln N_a - \ln N_a - (N_A \ln N_A + N_B \ln N_B)\}$$

$$= - R\,(N_A \ln N_A + N_B \ln N_B)$$

which is the same relationship as derived earlier by classical methods.

In a similar manner, many of the classical thermodynamic relationships and properties can be derived by statistical techniques. Thus, the macro thermodynamic properties are derivable from statistical concepts.

In addition, many thermodynamic properties can be estimated taking clues from the molecular properties and applying concepts from statistical thermodynamics.

Chapter- 10

Reaction Kinetics

The thermodynamic properties, especially the free energy change of a reaction, establishes whether a particular reaction is feasible or not, but it does not indicate at what rate the reaction would takes place. To enhance the rates of chemical reaction, increasingly higher temperatures have been employed over the years, as in the case of extraction of iron, and it has resulted in greater success of the indirect processes over the more direct sponge iron making processes. But persistent attempt at enhancing the rate of solid state reduction process has led to re-emergence and commercial success of the lower temperature and solid state reduction processes, such as the sponge iron making process. Although the attempts at enhancing the rate of solid state reduction has largely been by trial and error, but the science of the study of Chemical Reaction Kinetics (or simply Reaction Kinetics) has been a significant contributor. For further enhancement of rates, Reaction Kinetics is expected to play a key role in future.

Normally the subject of Reaction Kinetics would not form part of a monograph on Thermodynamics. However, there is another school of thought, which says that thermodynamics and kinetics are inseparable. Kinetics alone cannot be understood properly without any understanding of chemical thermodynamics, as can be seen from the discussions later in this chapter. The main purpose of inclusion of a section on reaction kinetics has been to give an easy introduction of this subject to the reader. Most of the books on Reaction Kinetics get into involved aspects of the subject right from the beginning. This section has been meant to give an initial and easy understanding of this subject. Another reason why we have decided to include this section is that, even though kinetics decides the rate of reaction, the driving force is decided by thermodynamics; for example, the free energy change, the equilibrium constant or the partial pressure. Any study of the rate of real process has to take into account both the thermodynamic properties and the kinetic factors.

10.1. FACTORS AFFECTING REACTION RATES

As implied earlier, the science of Reaction Kinetics studies rates of chemical reaction. Chemical reaction rates (meaning mass of reactants transformed per unit time) generally depend on the following:

- Concentrations (or more accurately the thermodynamic activities) of reactants
- Temperature of reaction – increase in temperature increases reaction rate
- Physical state of reactants – whether in a gaseous phase or in a solution (homogeneous reactions) or in separate phases (heterogeneous reaction as in a sponge iron rotary kiln) – whether in solid blocks (low surface area) or in powder, colloidal or emulsion form (high surface area). Higher the surface area, higher are the reaction sites and therefore higher are the reaction rates.
- Presence of Catalyst or Inhibitor
- Means of excitation – Light, Ultraviolet or Radioactive Radiation, etc.

10.2. RATE LAW AND ORDER OF REACTION

The most simple reaction is where one constituent breaks up into a number of product constituents; *e.g.*,

$$A = B + C + D + \ldots\ldots$$

If there is no appreciable reverse reaction then rate of reaction is represented by $-d\,[A]/dt$, where parentheses '[]' represents concentration of the constituent A. The sign is negative because the concentration of A is decreasing.

'Rate Law' states that the rate of reaction is proportional to the concentration of the reactant(s). Hence,

$$-\,d\,[A]/dt = k\,[A]$$

k is referred to as the Velocity Constant of reaction

or $$d\,[A]/[A] = -\,kdt$$

Integrating we get

$$\ln\,[A\,] = -\,kt + C$$

If, in the beginning (*i.e.*, when $t = 0$), $[A] = [A]_0$,

then, $$C = \ln\,[A]_0$$

and $$\ln\,([A]/[A]_0) = -\,kt$$

or $$[A]/[A]_0 = e^{-kt}$$

If, in time t, x amount of A has reacted, then,

$$[A] = [A]_0 - x$$

and $$[A]_0 - x = [A]_0\,e^{-kt}$$

or $$x = [A]_0\,(1 - e^{-kt})$$

On the other hand,

$$k = (1/t)\ln\,(\,[A]_0\,/\,[A]\,) = -\,(2.303/t)\log\,\{\,([A]_0 - x)/[A]_0\,\}$$

$$= -\ (2.303/t)\ \log\ (1 - x/[A]_0\,)\ )$$

A plot of $-\log(1 - x/[A]_0)$ against time t would be a straight line with a slope of $k/2.303$ from which value of k can be calculated. This particular relationship is said to hold for what is called a "First Order Reaction."

10.3. BIRTH OF THE RATE LAW OR LAW OF MASS ACTION

In 1799, Proust propounded the Law of Constant Proportions which stated that 'elements combine in definite proportions by weight and the composition of a pure chemical compound thus formed is independent of the method by which it is prepared.' Bertholet opposed this law bitterly and spent next nearly 60 years trying to bring evidences against it only to be rebutted one by one. In his quest for gathering such evidences, it finally dawned on him and his co-workers that he had been bringing up evidences of partial combination and the extent of combination, in some way, depended on the concentrations of the reactants used. From these results, Guldberg and Waage in 1864 deduced the Law of Mass Action, also known as the Rate Law. This law stated that "the rate at which a substance reacts is proportional to its active mass," and "the velocity of a chemical reaction is proportional to the product of the 'active masses' of the reactants." The active mass was assumed to be proportional to the molar concentration, as has been used in derivation in the last section.

10.4. UNIMOLECULAR REACTION & FIRST ORDER REACTION

In the derivation of section 10.2 we have taken the example of a unimolecular reaction, *i.e.*, a single molecule of A breaks into a number of molecules. We have also said that the final derivation

$$-\log(1 - x/[A]_0) = (k/2.303)\ t$$

is applicable to a first order reaction.

By a similar logic, for a bimolecular reaction of the type

$$A + B = C + D + \ldots\ldots.$$

and starting with equal concentrations of A and B (*i.e.*, $[A]_0 = [B]_0$) we get the relationship

$$1/(1 - x/[A]_0) = 1 + akt$$

and this applies to a second order reaction.

Hence, although the concept of 'Order of Reaction' has evolved from the molecularity of reaction, but they need not be the same. While molecularity refers to the number of molecules involved in a chemical reaction, the order of reaction must be derived from the study of the variation of reaction rates with time, concentration and (as we shall see later) temperature.

10.5. DECOMPOSITION OF AN OXIDE

Let us consider a chemical reaction where an oxide breaks into its elements.

$$MO = M + ½O_2$$

Degree of reduction, in such a case, is defined by the ratio of "oxygen removed from the oxide" to the "total removable oxygen." and is represented by the symbol α. Thus,

$$\alpha = 1 - [MO]/[MO]_o$$

and a plot of – log(1 – α) with *t* at any given temperatures should give a straight line with a slope of *k*/2.303 from which the value of velocity constant of reaction at that temperature can be calculated. This way the reaction is found to be of the first order. But if the plot is not a straight line, the reaction is not of first order, even though it remains a unimolecular reaction. Very often the reactions take place in a number of steps and are much more complex than what they appear at first glance. We shall see later that the reduction of iron oxide to metallic iron in a rotary kiln is also highly complex and the actual mechanism of reaction can only be conjectured at the moment.

10.5.1. Degree of Reduction

We have defined degree of reduction in the last section. If we take that definition literally for different types of iron oxides (such as hematite and magnetite) it would become highly difficult to compare results from different sets of experiments.

For circumventing this problem, we define ∝ = 0 for that iron oxide where all iron atoms are present as Fe_2O_3 molecules. If we have to start a study with a sample of, say, magnetite *i.e.*, we say that the sample, to start with, is already 11.11% reduced and ∝ = 0.1111. It becomes easier to compare results of different studies while using such a basis for defining degree of reduction.

10.6. TEMPERATURE DEPENDENCE OF VELOCITY CONSTANT: THE ARRHENIUS EQUATION

We have so far dealt with velocity constants of reactions at a particular temperature. In 1887 Arrhenius postulated that the velocity constant *k* of chemical reactions change with temperature as per the relationship

$$k = Ce^{-E/RT}$$

E is referred as the activation energy and it can be visualised as a barrier (an energy barrier) to the chemical reaction (Fig.10.1). Reacting molecules must acquire, during random collisions, interactions or excitation, enough energy so that it is able to cross the dam-like barrier and form the product. An increase in temperature would mean that molecules in general have acquired higher energy and there is an increased probability of greater number of molecules crossing that barrier in a given time.

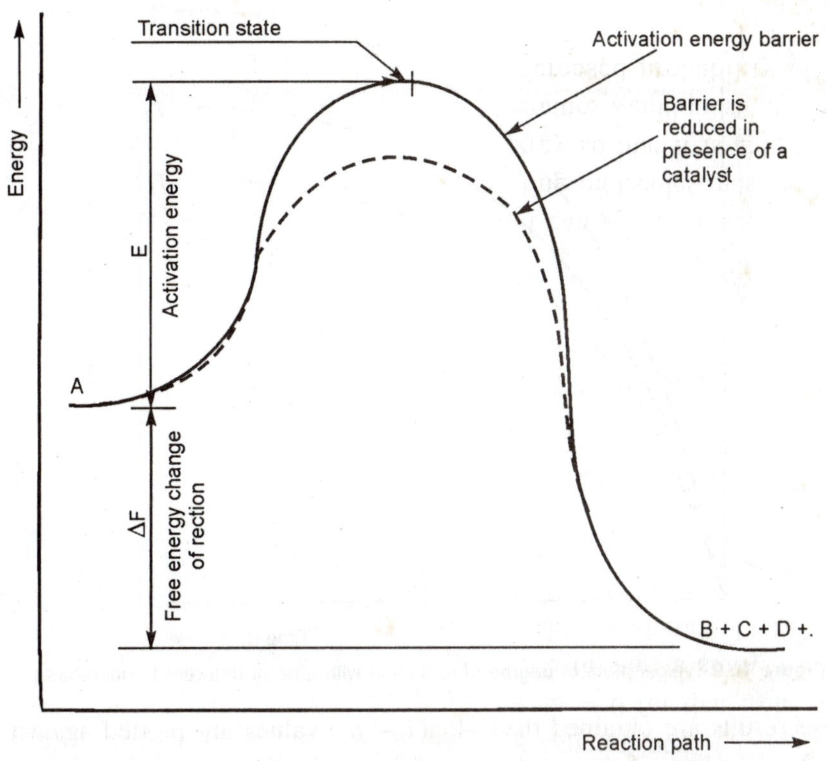

Figure 10.1 Schematic diagram representing activation energy barrier and the effect of a catalyst

Thus,

$$\ln k = C' - E/RT$$

or

$$\log k = C'' - (E/4.576)(1/T)$$

where R and E are expressed in calories. Thus, for evaluating E, we need to plot $\log k$ against $1/T$ and measure the slope of the resulting straight line.

10.7. EXPERIMENTAL DETERMINATION OF ACTIVATION ENERGY

In case of decomposition of an oxide as described in section 10.5, plots of degree of reduction against time would be of the type given in Fig. 10.2. These plots for different temperatures are experimentally determined. Various experimental techniques are followed to obtain the desired results, but Thermogravimetry is a quick and convenient method if adequate precautions are taken.

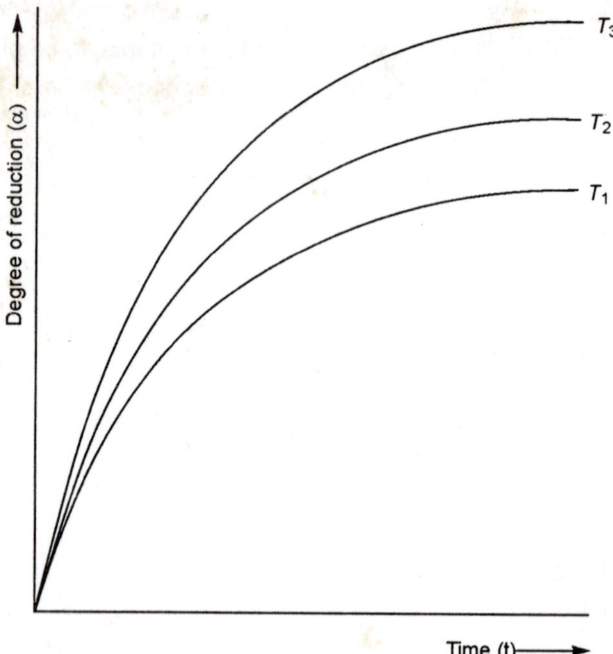

Figure 10.2 Typical plots of degree of reduction with time at different temperatures

Once these results are obtained then $-\log(1 - \alpha)$ values are plotted against time t, and they should be straight lines if the reaction is of first order. If reasonable linearity is observed then their slopes are measured (Fig. 10.3) to obtain k or velocity constant values. Thereafter, $\log k$ values are plotted against $1/T$ (reciprocal of absolute temperatures), and slope of this plot is a measure of activation energy. In fact, in caloric terms, the slope is equal to $-E/4.576$, where E is the activation energy (Fig. 10.4).

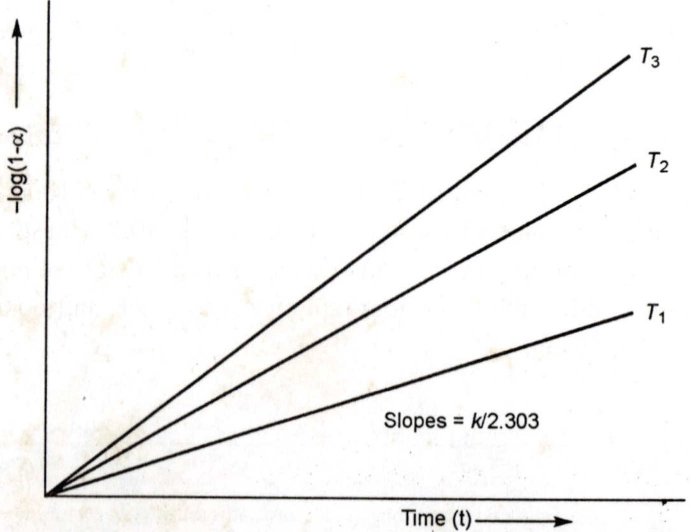

Figure 10.3 Plot of $-\log(1-\alpha)$ with time (straight lines denote first order reaction)

If plots in Fig. 10.3 are not straight lines the reaction is not of first order. For second order reaction, instead of – log (1 – α), a different function of α needs to be plotted along the Y-axis. This function is 1/(1 – α). This time the plots would be straight lines, but not passing through the origin.

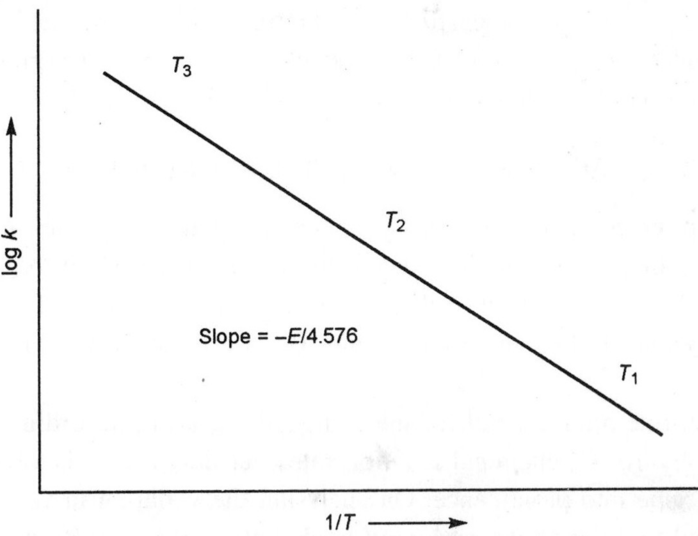

Figure 10.4 Plot of log of velocity constant against reciprocal of absolute temperature (Arrhenius Plot)

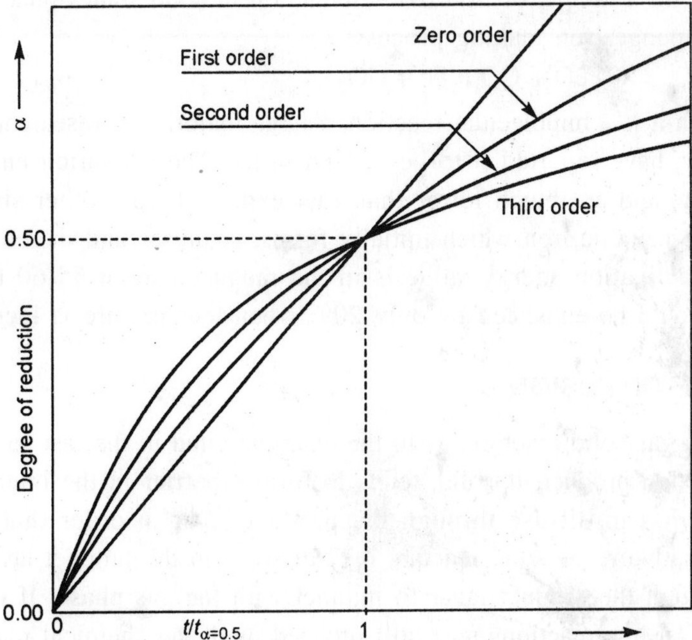

Figure 10.5 Reduced time plots for different orders of reaction

Instead of trying to find out the order of reaction by trial and error, a short cut is normally applied by using what is called a reduced time plot. In this plot $\propto$ is plotted against a dimensionless time, which is obtained by taking ratio of the actual time with the time required to attain $\alpha = 0.5$. On the plot obtained experimentally, theoretical plots of reaction by different mechanisms are superimposed and then the matching with a particular mechanism can be quickly observed. Many times in a particular reduction region, one mechanism may be followed, which would be giving way to another mechanism in another region, and this also can be deduced by comparing with standard plots (Fig. 10.5).

10.8. VARIATION OF REACTION RATE WITH TEMPERATURE

The treatment presented in the proceeding sections is applicable to the rates of reactions. First order and second order reactions have been discussed. Reaction can be of higher order (third order and above) as also of fractional order. Reactions can also be of zero order where there is no effect of concentration on reaction rate. But we hardly ever come across such a reaction.

There is a thumb-rule often quoted for metallurgical reactions, according to which if we increase temperature by 10 °C, chemical reaction rates get doubled. It is very unclear as to how this thumb-rule came into prominence. On analysing the variation in velocity constant at about 1000 °C we find that the above reasoning holds only if the activation energy is of the order of 220 kCal/mole.

Let us take the example of iron oxide reduction in solid state by carbon monoxide, as it happens in the sponge iron making process.

$$FeO + CO = Fe + CO_2$$

Even though it is a bimolecular reaction, a large majority of researchers who investigated it experimentally, have reported it to be of first order. The activation energy values though, vary very widely, and are much lower than that expected from other similar reaction. It is reasoned that the metallic iron which initially forms, catalyses and therefore enhances further reaction. If the activation energy value is in the range of about 55-60 kCal/mole, then the reduction rate would be enhanced by only 20% when temperature is increased by 10 °C.

10.9. ROLE OF DIFFUSION

In case of a gas-solid reaction, as in the example cited in the last section on sponge iron making, the reaction product, if solid, tends to form a barrier in the interaction process. The gaseous species has to diffuse through the product layer in order that further reaction is continued. Alternatively, the solid reactant may dissolve in the product layer, or in some other way diffuse through the product layer to interact with the gas phase. If these diffusion rates are sufficiently fast, the reactions may still proceed, with the chemical reaction rate being the controlling step, and activation energy of the chemical reaction would determine the variation of reaction rate with temperature. This probably happens in the sponge iron making process as a majority of researchers have reported this process of sponge iron making to be "Chemical Reaction Rate Controlled."

But if diffusion rates are too slow the activation energy of chemical reaction would not have any role in the progress of reaction and we would have a situation where the progress of reaction is "Diffusion Controlled." In intervening regions where both chemical reaction rate and rates of diffusion have influence on overall progress of reaction we say that the reaction progress is in the "Mixed Control" domain.

We would not discuss here the diffusion mechanisms and diffusion controlled processes. The treatment in complex and other comprehensive documents may be referred for this purpose. We would only emphasise the need to find means to circumvent diffusion steps in reactions involving condensed phases (either solids or liquids) as diffusion barriers often make industrial exploitation of process unviable.

10.10. HOW SPONGE IRON MAKING IS A FIRST ORDER REACTION?

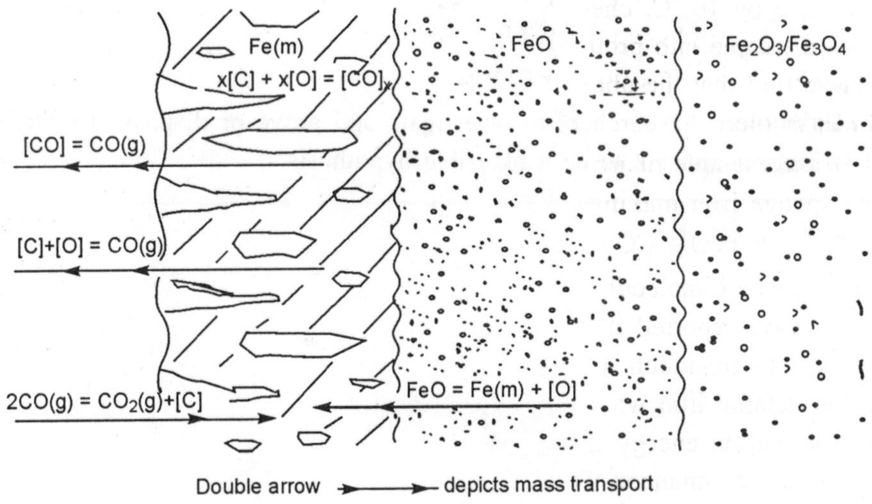

Figure 10.6 A possible mechanism of pore formation during reduction of Iron Oxide

As mentioned earlier that although the following reaction

$$FeO + CO = Fe + CO_2$$

is bimolecular, a very large number of researchers have found it to be of first order. A few reasons may be responsible for this observation.

- FeO being in condensed phase, its activity (representing concentration) does not change with time and therefore only the concentration of CO is affecting the rate of reduction.

- The reaction is, in some way, being catalysed. Many say that the catalysis is by the iron metal, which is produced in the reaction. During catalysis, an intermediate single molecule is formed, which then breaks into the products. Therefore, the reaction is of the first order.

Largely following the second reasoning, a hypothesis on the mechanism of reduction has been presented in Fig. 10.6. Since there is large volume reduction (over 50%) during sponge iron making, a lot of pores form on the product layer, which is formed by the reduction on surface of iron ore lump or pellet. While these pores facilitate easier access to reducing gases, it is not necessary for CO molecules to actually meet FeO molecules to effect further reduction. Carbon and oxygen have significant solubilities in metallic iron. The dissolved carbon and oxygen atoms within the metal layer probably form an associated molecule (CO dissolved in metallic iron), which get liberated on the surface to give CO gas and thereby effecting the reduction of FeO.

$$2CO(g) = CO_2(g) + [C]$$

Here parentheses '[]' mean that the constituent is dissolved in Fe_m layer.

$$FeO(s) = [Fe] + [O]$$

$$x[C] + x[O] = [CO]_x$$

$$[CO]_x = x\ CO(g)$$

It is open to future researchers to investigate and prove or disprove this hypothesis. At the time of writing, it appears to be a plausible hypothesis.

Definitions of Selected Terms

1.	Isothermal Process	:	Constant temperature process	$dT = 0$
2.	Isochoric Process	:	Constant volume process	$dV = 0$
3.	Isobaric Process	:	Constant pressure process	$dP = 0$
4.	Adiabatic Process	:	Constant heat process	$dQ = 0$
5.	Enthalpy (H)	:	$H = E + PV$	
6.	Entropy (S)	:	$dS = dQ_{rev}/T$	
7.	Work Function, Helmholtz Free Energy (A)	:	$A = E - TS$	
8.	Free Energy, Gibbs Free Energy (F)	:	$F = H - TS$	

Epilogue

An expert reader will surely note many shortcomings of the book if he is keen on a comprehensive text. The authors, however, had a different objective. They did not include detailed discussions of all topics because they wanted to present a 'comprehensible' text and not a 'comprehensive' one. There are many books on metallurgical thermodynamics already and there may not be need for another comprehensive textbook.

Thermodynamics has evolved out of inspired mathematical reasoning and critical observations of some common phenomena. To the student who is first introduced to the subject, many existing textbooks make the subject complicated and formidable. The authors believe that there is need to first make the student familiar with the fundamental concepts in a simple way without sacrificing scientific rigour.

For centuries, many chemical and metallurgical processes were developed essentially through trial and error. Industries held valuable secrets discovered by their operators after years of experience. Those processes, however, were better understood and refined using the understanding of basic concepts of free energies, activity coefficients, interaction parameters etc. In recent decades, however, newer processes have been developed more on theoretical calculations to start with rather than trial and error. An atomic reactor cannot be built by trial and error.

Grasp of theory needs, first, the understanding of the basic concepts and principles. The present text is an attempt to help the reader understand the concepts quickly and easily. Those who have been exposed to the subject of thermodynamics can recapitulate their knowledge and study the text to clear doubts.

The authors would be happy if the text serves the purpose mentioned and would welcome comments and suggestions from the readers regarding mistakes, ambiguities and any other shortcomings.

HSR

Bibliography

- Abbot, M.M. and Van Ness. H.C., Theory and Problems of Thermodynamics, McGraw-Hill International Book Co., Singapore (1981).
- Beattie, James A. & Irwin Oppenheim, Principles of Thermodynamics, Elsevier Scientific Publishing Co., Amsterdam (1979).
- Benson, Rowland S., Advanced Engineering Thermodynamics, 2nd edition, Pergamon Press, Oxford (1977).
- Bhaskaran, K.A. & Venkatesh A., Problems in Engineering Thermodynamics, Tata McGraw-Hill Publishing Co. Ltd., New Delhi (1978).
- Bodsworth, C. & Bell H.B., Physical Chemistry of Iron & Steel Manufacture, 2nd edition, Longmans, London (1972).
- Campbell, A.N. & Smith N.O., The Phase Rule and its Applications, Dover Publication Inc. (1951).
- Carnot, S.; Quoted in 'The Second Law of Thermodynamics' by Magie W.F.(Ed.), Harper and Brothers, New York (1899).
- Christensen, James J., Hanks R.W. & Izatt R.N., Handbook of Heats of Mixing, John Wiley & Sons, New York (1982).
- Clausius, R.; Quoted in 'The Second Law of Thermodynamics,' Magie W.F. (Ed.), Harper and Brothers, New York (1899).
- Coudurier, Lucien, Donald D. Hopkins & Igor Wilkomirsky, Fundamentals of Metallurgical Processes, 2nd edition, Pergamon Press, Oxford (1985).
- Darken, Lawrence S. & R.W. Gurry, Physical Chemistry of Metals, McGraw-Hill Book Co., New York (1958).
- Devereux, Owen F., Topics in Metallurgical Thermodynamics, John Wiley & Sons (1983).
- Elliot, J.F. & Meadowcraft T.R., Ed., Steelmaking – The Chipman Conference, MIT Press, Massachusetts (1965).
- Elliott, J.F., Gleiser M. & Ramakrishna V., Thermochemistry for Steelmaking (2 volumes), Addison Wesley Publication Co., Reading, Massachusetts (1960/63).
- Engell, H.J., Z. Phys. Chem. (Frankfurt), **35**, 192-5 (1962).
- Fermi, Enrico, Thermodynamics, Courier Dover Publications, New York (1956).

- Fisher, R.M., Oriani R.A. & Turkdogan E.T., Physical Chemistry in Metallurgy, Proceedings of the Darken Conference, United States Steel (1976).
- Foster, P.K. & Welch A.J.E., Trans. Faraday Soc., **52**, 1636-42 (1956).
- Gaskell, D.R., Introduction to Thermodynamics of Materials, 4th edition, Taylor & Francis, Washington DC (2003).
- Ghosh, Ahindra, Textbook of Materials and Metallurgical Thermodynamics, Prentice Hall of India, New Delhi (2003)
- Glasstone, Samuel, Thermodynamics for Chemists, D. Van Nostrand Co. Inc., New York (1947).
- Haase, Rolf, Thermodynamics of Irreversible Processes, Addison-Wesley Publishing Co., Reading, Massachusetts (1960).
- Holman, J.P., Thermodynamics, 3rd edition, McGraw-Hill Kogakusha Ltd., Tokyo (1980).
- Hsieh, Jui Sheng, Principles of Thermodynamics, Scripta Book Co., Washington D.C./ McGraw-Hill Book Co., New York (1975).
- Hultgren Ralph, Desai P.D., Hawkins D.T., Gleiser M., Kelly K.K. & Wagman D.D., Selected Values of Thermodynamic Properties of Elements & Binary Alloys, American Society for Metals, Ohio (1973).
- Infelta, Pierce, Introductory Thermodynamics, Brown Walker Press, Florida (2004).
- Jacob, K.T., G.M. Kale & K.P. Abraham, J. Electrochem. Soc., **139** (2), 517-520 (February 1992).
- Jorgensen, Christian and Ingegert Thorngren, Thermodynamic Tables for Process Metallurgists, Almqvist & Wiksell, Stockholm (1969).
- Jost, Wilhelm, (Ed.), Physical Chemistry: An Advanced Treatise – Vol 1: Thermodynamics, Academic Press, New York (1971).
- Joule, J.P., The Scientific Papers of James Prescott Joule, The Physical Society, London (1884).
- Kapoor, Madan Lal, Chemicals and Metallurgical Thermodynamics, Nem Chand & Bros., Roorkee, India (1981).
- Karapetyants, M.Kh., Chemical Thermodynamics, Mir Publishers, Moscow (1978).
- Kelvin, Lord William Thomson, Philosophical Magazine, October 1848, Vol. 1, Cambridge University Press (1882), pp. 100-106.
- Kenney, W.F., Energy Conservation in Process Industries (1984).
- Khangaonkar, P.R., Metallurgical Thermodynamics, NEBS Publishers & Distributors, Nagpur, India (1967).
- Knuth, Eldon L., Introduction to Statistical Thermodynamics, McGraw-Hill Book Co., New York (1966).
- Kubaschewski, O. & C.B. Alcock, Metallurgical Thermochemistry, 5th edition, Pergamon Press (1979).
- Lewis, Bernard & Guenther von Elbe, Combustion, Flames and Explosions of Gases, Academic Press Inc., Orlando (1987).
- Lewis, Gilbert Newton & Merle Randall, Thermodynamics, McGraw-Hill Book Co., New York (1961).

- Lupin, Claude H.P., Chemical Thermodynamics of Materials, Prentice Hall (1983).
- Moran, M.J., Fundamentals of Engineering Thermodynamics.
- Morrill, Bernard, An Introduction to Equilibrium Thermodynamics, Pergamon Press Inc. (1972).
- Nernst, W., Experimental and Theoretical Applications of Thermodynamics to Chemistry, C. Scribner's Sons, New York (1907).
- Ness, H.C. Van, Understanding Thermodynamics.
- Obert, Edward F., Elements of Thermodynamics and Heat Transfer, McGraw Hill Book Co., Inc., New York (1949).
- Plank, M., Treatise on Thermodynamics, translated by A. Ogg, Longmans Green and Co., London (1927).
- Prasad, K.K., Ph.D. Thesis, Indian Institute of Science, Bangalore (1971).
- Prasad, K.K. & Abraham K.P., Proc. Symp. Materials Science Research, Bangalore, **2**, 127-33 (1970).
- Prasad, K.K. & Abraham K.P., Trans. Indian Inst. Metals, **27**(4), 259-61 (1974).
- Prasad, K.K., Raghavan S. & Abraham K.P., Proc. Symp. Phase Transformations & Phase Equilibria, Bangalore, 105-118 (1975).
- Prasad, K.K., Seetharaman S. & Abraham K.P., Trans. Indian Inst. Metals, **22**(4), 7-9 (1969).
- Raghavan, S., Hajra J.P., Iyengar G.N.K.& Abraham K.P., Proc. Symp. Phase Transformations & Phase Equilibria, Bangalore, 119-132 (1975).
- Rao, Y.K., Stoichiometry and Thermodynamics of Metallurgical Processes, Cambridge University Press, Cambridge (1985).
- Raznjevic, Kuzman, Handbook of Thermodynamic Tables and Charts, Hemisphere Publishing Corporation, Washington/McGraw-Hill Book Co., New York (1976).
- Richardson, F.D., Physical Chemistry of Melts in Metallurgy (2 Vols.), Academic Press, New York (1974).
- Rock, Peter A., Chemical Thermodynamics: Principles and Applications, The Macmillan Co., Ontario (1969).
- Rogers, G.F.C. & Mayhew Y.R., Engineering Thermodynamics, Longman Group Limited, London (1980).
- Rosenberg, Robert M., Principles of Physical Chemistry, Oxford University Press, New York (1977).
- Roy, Bimalendu Narayan, Fundamentals of Classical and Statistical Thermodynamics, John Wiley & Sons, New York (2002).
- Sage, Bruce H., Thermodynamics of Multicomponent System, Reinhold Publishing Corporation, New York (1965).
- Sandler, Stanley I., Chemical and Engineering Thermodynamics, John Wiley & Sons, New York (1977).
- Schenk, R., Franz H. & Willeke H., Anorg. Allg. Chemie, **184**, 1-38 (1929).

- Seetharaman, S., Ph.D. Thesis, Indian Institute of Science, Bangalore (1970).
- Seetharaman, S. & K.P. Abraham, Indian Jl. Technology, **6**(4), 123 (1968).
- Shah, A.K., Prasad K.K. & Abraham K.P. , Trans. Indian Inst. Metals, **24**(1), 40-2 (1971).
- Swerdtfeger, Muan K. A, Trans. Met. Soc. AIME, **239**(8), 1114-9 (1967).
- Smith, J.M. & Van Ness H.C., Introduction to Chemical Engineering Thermodynamics, 3rd edition, McGraw-Hill Book Co., New York (1975).
- Sonntag, Richard E., Clans Borgnakke & Gordon J. Van Wylen, Fundamentals of Thermodynamics.
- Thomson, W., Mathematical and Physical Papers, University Press Cambridge (1882).
- Tien, Chang L. & John H. Lienhard, Statistical Thermodynamics, Holt, Reinhart and Winston Inc., New York (1971).
- Tupkary, R.H., Metallurgical Thermodynamics, T.U. Publisher, Nagpur, India (1995).
- Turkdogan, E.T., Physical Chemistry of High Temperature Technology, Academic Press, New York (1980).
- Upadhyaya, G.S. & Dube R.K., Problems in Metallurgical Thermodynamics and Kinetics, Pergamon Press, Oxford (1977).
- Wagner, C., The Thermodynamics of Alloys, Addison-Wesley Press, Cambridge (1952).
- Ward, R.G., An Introduction to the Physical Chemistry of Iron & Steel Making, ELBS/ Edward Arnold Publishers, London (1965).
- Wark, Kenneth, Thermodynamics, 3rd edition, McGraw-Hill Book Co., New York (1977).
- Wartenberg, Von H.V., Reusch H.J. & Saran E., Z. Anorg. Allgem. Chemice, **230**, 257-76 (1937).
- Weber, Harold C., & Herman P. Meisner, Thermodynamics for Chemical Engineers, Wiley Eastern (P) Ltd., New Delhi (1970).
- Wylen, Gordon J. Van, Thermodynamics, John Wiley & Sons Inc. (1959).
- Zemansky, M.V. & Dittman R.H., Heat and Thermodynamics (1981).

Subject and Key Word Index